THE EFFECT OF THE DEMOGRAPHICS OF INDIVIDUAL HOUSEHOLDS ON THEIR TELEPHONE USAGE

THE EFFECT OF THE DEMOGRAPHICS OF INDIVIDUAL HOUSEHOLDS ON THEIR TELEPHONE USAGE

BELINDA B. BRANDON, editor

BALLINGER PUBLISHING COMPANY
Cambridge, Massachusetts
A Subsidiary of Harper & Row, Publishers, Inc.

Copyright © 1981 by Bell Telephone Laboratories, Incorporated. All rights reserved. No part of this publication may be reproduced, stored in a retrieval system, or transmitted in any form or by any means, electronic, mechanical, photocopy, recording or otherwise, without the prior written consent of the publisher.

International Standard Book Number: 0–88410–695–0

Library of Congress Catalog Card Number: 80–27158

Printed in the United States of America

Library of Congress Cataloging in Publication Data

Main entry under title:

The Effect of the demographics of individual households on their telephone usage.

  Bibliography: p.
  Includes index.
  1. Telephone–Illinois–Chicago–Addresses, essays, lectures. I. Brandon, Belinda B.
HE8841.C53E33   384.6'09773'11   80–27158
ISBN 0–88410–695–0

DEDICATION

In memory of
Marcia S. Bell

CONTENTS

List of Exhibits xiii

Foreword xv
Belinda B. Brandon, Edmund C. Hoeppner, and Lawrence Garfinkel

Chapter 1 1
Introduction and Summary
Belinda B. Brandon and Paul S. Brandon

1.1	Motivation for the Study	1
1.2	The Rate Structure of Telephone Calling for Chicago Customers	4
1.3	Highlights of the Results Relating Telephone Usage with Demographic Characteristics	5
	1.3.1 Aggregate Local and Aggregate Suburban Usage	5
	1.3.2 Local and Suburban Usage by Time of Day	7
	1.3.3 Local and Suburban Usage by Distance	8
	1.3.4 Total Charges and Vertical Service Charges	8
	1.3.5 Toll Usage	9
	1.3.6 Remarks on the Empirical Results	9
1.4	Observations on the Impact of Pricing Changes	9
1.5	The Regulatory Dilemma	11
1.6	Benefits and Limitations of the Study	13
1.7	A Methodological Overview	14
1.8	Preview	18

Chapter 2 19
The Development of the Customer Sample
Wm. H. Williams and Michael L. Goodman

2.1	Introduction	19
2.2	Relevant Characteristics of Telephone Equipment in the City of Chicago	19
2.3	The Sample	20
	2.3.1 Easy Equipment	20

2.3.2 The Sample Draw	21
2.4 Some Demography of Chicago	21
Editor's Note Regarding Inferences Beyond the Sample	26
2.A Appendix: The Calculations for Exhibit 2.2	27

Chapter 3   29
The Questionnaire, Calling, and Billing Data
Belinda B. Brandon

3.1 Introduction	29
3.2 Questionnaire Design	30
3.3 Questionnaire Administration	33
3.4 Response Rate	35
3.5 Error Checking	38
3.6 The Calling and Billing Data	41
3.7 Conclusion	45
3.A Appendix: Mailed Version of the Questionnaire	46

Chapter 4   57
The Data Base Design and Implementation
Robert McGill

4.1 Introduction	57
4.2 Overview of the Data	57
4.3 Order From Chaos — Imposing Structure	58
4.4 Users and Anticipated Use	61
4.5 Media, Methods, and Efficiency	62
4.6 Bringing up the Data Base	63
4.7 The Data Base as Seen by the User	65
4.8 Physical Description of the Data Base	67
4.9 Additional Data	70
4.10 With Perfect 20/20 Hindsight	71
4.11 Conclusions	72
4.A Appendix: Random and Sequential Data Access	73

Chapter 5   75
Box Plot Analysis of Local and Suburban Calling Frequencies and Conversation Times
*Belinda B. Brandon, Elsa M. Ancmon,
Carole C. Finck, and Robert H. Groff*

5.1 Introduction	75

5.2	The Data	76
5.3	Tutorial on the Notched Box Plot	77
5.4	Local Calls	79
	5.4.1 Numbers of Local Calls	81
	5.4.2 Removal of Rounding Bias from Durations	82
	5.4.3 Average Durations of Local Calls	85
	5.4.4 Total Local Conversation Times	85
	5.4.5 Customer-Estimated vs Observed Numbers and Durations of Local Calls	86
5.5	Suburban Calls	87
	5.5.1 Numbers of Suburban Calls	88
	5.5.2 Average Durations of Suburban Calls	88
	5.5.3 Total Suburban Conversation Times	89
	5.5.4 Suburban Message Units	89
5.6	Total Local and Suburban Message Units	90
5.7	Usage by Message-Unit Allowance	91
5.8	Summary and a Concluding Note	92
5.A	Appendix: Wilcoxon-Mann-Whitney Rank Sum Tests	93
5.B	Appendix: Box Plots of Local and Suburban Usage by Demographic Group	95

Chapter 6
Regression Analysis of Local and Suburban Usage
Paul S. Brandon

133

6.1	Introduction	133
6.2	The Data	135
6.3	The Econometric Model and Estimation Procedure	141
6.4	The Results	151
	6.4.1 Number of Local Calls	156
	6.4.2 Average Duration of Local Calls	156
	6.4.3 Total Local Conversation Time	157
	6.4.4 Suburban Message Units	157
	6.4.5 Total Message Units	157
6.5	Conclusion and Suggestions for Future Research	158
6.A	Appendix: Illustration of Generalized Least Squares Procedure	159

Chapter 7
The Effects of Time of Day on Local and Suburban Calling Frequencies and Conversation Times
Belinda B. Brandon, Paul S. Brandon, and Elsa M. Ancmon

165

7.1		Introduction	165
7.2		The Data	167
7.3		Results by Rate Periods	167
	7.3.1	Numbers of Local Calls by Rate Periods	169
	7.3.2	Average Durations of Local Calls by Rate Periods	170
	7.3.3	Total Local Conversation Times by Rate Periods	171
	7.3.4	Suburban Calling by Rate Periods	171
	7.3.5	Local and Suburban Message Units	173
7.4		Results by Hours of the Day	173
	7.4.1	Numbers of Local Calls by Hours of the Day	174
	7.4.2	Mean Durations of Local Calls by Hours of the Day	176
	7.4.3	Total Local Conversation Times by Hours of the Day	177
	7.4.4	Suburban Calling by Hours of the Day	178
	7.4.5	Total Local and Suburban Message Units by Hours of the Day	179
7.5		Conclusions and Suggestions for Future Research	180
7.A		Appendix: Rank Sum Tests for Differences Between Demographic Groups by Rate Periods	182
7.B		Appendix: Local and Suburban Usage by Demographic Group: Box Plots by Rate Period and Means by Hour of the Day	183
7.C		Appendix: Mean Usage of Demographic Groups Aggregated Over Weekdays and Over Weekends	216

Chapter 8 219
The Association of Distance with Calling Frequencies and Conversation Times
Belinda B. Brandon and Elsa M. Ancmon

8.1		Introduction	219
8.2		The Data	220
8.3		Calculation of Distance and Specification of Proximity	221
8.4		The Geographic Representativeness of the Sample	223
8.5		Central Office Proximity Results	224
8.6		Distance Results	225
	8.6.1	Number of Local and Suburban Calls	227
	8.6.2	Average Durations of Local and Suburban Calls	228
	8.6.3	Total Local and Suburban Conversation Time	228
	8.6.4	Total Local and Suburban Message Units	229
8.7		Aggregate Measures of Local and Suburban Calling	229
8.8		Distance Results by Time of Day	231

	8.8.1 Average Distance by Time of Day	232
	8.8.2 Total Minute-Miles by Time of Day	232
	8.8.3 Average Minute-Miles by Time of Day	233
8.9	Conclusion	233
8.A	Appendix: Local and Suburban Calling by Distance	234
8.B	Appendix: Mean of Combined Local and Suburban Usage by Demographic Group	256
8.C	Appendix: Measures of Distance of Local and Suburban Calling by Time of Day	264

Chapter 9
Analysis of Billing Data
Robert H. Groff

275

9.1	Introduction	275
9.2	The Data	276
9.3	Methodology	277
	9.3.1 Preliminary Methodological Findings	277
	9.3.2 Computed Monthly Bill	278
	9.3.3 Vertical Service Charges	281
9.4	Regression Results	286
	9.4.1 Results for Computed Monthly Bill	286
	9.4.2 Results for Vertical Service Charge	288
9.5	Conclusions	291
9.A	Appendix: Plots for Computed Monthly Bills and for Vertical Service Charge	292
9.B	Appendix: Regression Results for Computed Monthly Bills and for Vertical Service Charge	297

Chapter 10
Toll Usage
Susan J. Devlin and I. Lester Patterson

303

10.1	Introduction	303
10.2	The Data	304
10.3	Description of the Toll Findings	305
	10.3.1 Number of Customers — Bar Graphs	305
	10.3.2 Number of Toll Calls	306
	10.3.3 Total Toll Conversation Time	309
	10.3.4 Average Duration of Toll Calls	310
	10.3.5 Toll Charges	311

10.4	Toll vs. Local and Suburban Calling — Demographic Comparisons	313
	10.4.1 Race	314
	10.4.2 Income	315
	10.4.3 Years of Residence in Chicago	316
	10.4.4 Age of the Head of Household	317
10.5	Conclusions and Recommendations	317
10.A	Appendix: Toll Calling by Demographic Group	320

Chapter 11 355
Other Studies and Suggestions for Future Research
Belinda B. Brandon, Paul S. Brandon, and Wm. H. Williams

11.1	Introduction	355
11.2	Other Studies	355
11.3	Suggested Sample Design for a Before-and-After Study	357
11.4	Questionnaire Administration and Design	362
	11.4.1 Questionnaire Administration	362
	11.4.2 Questionnaire Design	363
	11.4.3 Interviewer Questionnaire	364
11.5	Suggestions for Future Research	365
11.A	Appendix: Suggested Version of the Questionnaire for Mailing	367
11.B	Appendix: Suggested Version of the Questionnaire for Interviewer	377
11.C	Appendix: Interviewer Instructions, and Coding Instructions	390

Bibliography 395

Index 399

About the Editor 405

LIST OF EXHIBITS

2.1	Chicago Central Office Area Demographic Information	24
2.2	Comparison of Central Office Area Demography in Chicago	25
3.1	Completed Questionnaire Percentages	37
3.2	Monthly Usage Rates of Sampled Customers, With and Without Completed Questionnaires	38
3.3	Refusal Rates on Individual Questions	39
3.4	Amount of Data of Each Type for Each Date	42
3.5	Mean Monthly 1973 Telephone Usage of Sample Customers, With and Without 1974 Data	44
5.1	Standard Normal Probability Plots, Number of Local Calls	80
6.1	List and Type of Variables	137
6.2	Generalized Least Squares Regression Results	152
6.3	Standard Errors of Predicted Means	155
8.1	Chicago Sampled Central Offices	222
9.1	Customer Characteristics	280
9.2	Distribution of Latent and Actual Vertical Service Charges	284
11.1	When Data Should be Processed Near Rate Change	361

Foreword

This book is a collection of papers about a project that was designed to estimate how the use of telephone services of individual households is affected by characteristics such as income, family size, and race. The principal motivation for the project, as discussed more fully in Chapter 1, was to study local telephone usage, so as to understand better the effect on households of pricing local calls along cost-causing dimensions such as their number, duration, time of day, and distance. It is hoped that this documentation will facilitate the transmission of the methodologies used in the study as well as the study results themselves to telephone company managers, to regulatory authorities, and to other interested parties.

When the project was conceived by the editor in 1972, there were charges for local calls for both residence and business telephone customers in only a few cities in the United States — Boston, Chicago, Newark, and New York — and some of those had a flat-rate option. In addition to charging by the number of calls, Boston and Newark had charges according to the durations of calls. Illinois Bell Telephone Company had applied for the implementation of charging by local call durations in Chicago. For calls within New York City, New York Telephone Company was contemplating the initiation of charges by duration, time of day, and distance. Each of these, then, was a possible study site.

The Chicago site seemed particularly attractive on several counts. Firstly, the introduction of charging by durations was under active consideration by the Illinois Commerce Commission, and there seemed to be an excellent opportunity to study calling habits both before and after this change in price structure. Secondly, Chicago offered the possibility of a study that would be free of contaminants such as other simultaneous rate changes and options without duration charges. Thirdly, Illinois Bell Telephone Company had arranged to collect calling data for its own use from certain switching machines in Chicago, making it inexpensive to take data from those machines simultaneously for the present study. So the editor chose Chicago as the source of data.

Clearly the task of studying calling patterns is a large and difficult one and presents a problem to the researcher with finite resources. He may choose to pursue a very broad study with the danger of spreading resources too thin. Alternatively he may focus on too narrow a problem. In the study at hand, the editor decided to concentrate on the demand side of the problem, excluding an explicit study of costs, and, within demand, to focus on the residence rather than on the business market. The reasons for the choice of residence were two: Firstly, less is known about the impact of local usage charges on residence customers than on business customers because options for local usage charges are currently less prevalent for residence than for business customers. Secondly, the gathering of data for business seemed much more complex than for residence customers, since one must search out all the lines associated with a particular business customer.

The subsequent chapters document the methodology and the results of the study, written in each case by the principal investigators involved in the portion of the study reported upon. The reader may notice a variety of styles of presentation across chapters; they vary in technical difficulty, detail, degree of motivation, and other stylistic aspects. More importantly, there is considerable variation in the analytical methodology. We believe that this diversity is a strength, not a weakness. Different questions and different kinds of data call for different approaches. The variety of methodologies also has considerable pedagogical value. Thus the editor, while creating uniformity in terminology, has not discouraged this variety across chapters.

As is typical for an exploratory investigation, the annual manpower resources deployed in the study have been limited. For the first year or so after its beginning in 1972, they consisted primarily of the editor, a research economist at Bell Telephone Laboratories (BTL), who worked full time on the project with occasional consulting help from research statisticians from BTL and the Illinois Bell Telephone Company (IBT) and from experienced rate specialists at IBT and the American Telephone and Telegraph Company (AT&T). Later, other economists, statisticians, and programmers added their efforts to the study so that the typical input increased to between two and three equivalent full time researchers.

This study is primarily the work of economists and statisticians at BTL, where, by virtue of their administrative positions, technical and non-technical help was provided by Ramanathan Gnanadesikan, Henry O. Pollak, Frank W. Sinden, and Edward E. Zajac. Lawrence Garfinkel (AT&T), Edmund C. Hoeppner (IBT), James N. Kennedy (IBT), and numerous persons in their organizations provided assistance to those in

BTL who carried out the study. Wayne A. Larsen was heavily involved in the preliminary analyses of the data when he was at BTL. Technical consulting was also contributed by Richard A. Becker, Bruce C. N. Greenwald, Alan M. Gross, Jerry A. Hausman, Edward B. Fowlkes, William J. Infosino, Roger W. Klein, Roger W. Koenker, Colin L. Mallows, Carl Pavarini, Mary H. Shugard, Richard G. Stafford, William E. Taylor, Irma Terpenning, Paul A. Tukey, and Kenneth Wachter.

Several persons are thanked for their aid in individual chapters. But one person who should be praised here because she gave substantial clerical aid through almost the entire project is Stephanie Szabo. Strong appreciation is also due to Patti Martin, who typed into the computer almost all of the text of the book and prepared it for the phototypesetting system. Assistance in the phototypesetting process was also received from Elsa M. Ancmon, Paul S. Brandon, Robert McGill, and Stephanie Szabo.

>Belinda B. Brandon (Bell Telephone Laboratories)
>Edmund C. Hoeppner (Illinois Bell Telephone Company)
>Lawrence Garfinkel (American Telephone and Telegraph Company)

CHAPTER 1

Introduction and Summary

Belinda B. Brandon and Paul S. Brandon

1.1 Motivation for the Study

Regulatory authorities scrutinize and control the prices of goods and services in the public utility sector, including telecommunications. In the past they have emphasized control over the general level of each utility's prices, but in recent years they have become increasingly interested in controlling the structure of prices. For example, they are interested in whether the pricing of local telephone service should be unrelated to local usage or, alternatively, whether charges should be more reflective of the costs caused by the usage of different customers.

The regulatory problems presented by a major change in price structure are similar to those presented by a change in price level. Customers who feel adversely affected by a proposed pricing change may be sufficiently motivated to appear at a regulatory hearing to voice their objection. Those with similar objections may even organize into a group and retain legal counsel to plead their case.[1] In any event, policy makers — both regulators and telephone company executives — must

---
[1] There is a proposition in the economics literature that the probability that some group bears the costs of organizing and lobbying before the regulators is affected by the impact that a prospective decision would have on the group. For a formalization of this notion, see, *e.g.*, George J. Stigler, "The Theory of Economic Regulation", *Bell Journal of Economics and Management Science*, v. 2, no. 1 (Spring, 1971), pp. 3-21.

be prepared to respond responsibly to suggestions for and comments on proposed pricing changes. In addition, the regulator's personal prospects, budget, and so forth may be affected either directly or indirectly by the general satisfaction with their decisions on the part of their constituents. The regulators may also view particular groups as deserving, aside from any considerations of their political power.

Thus, both regulators and telephone company executives have expressed interest in knowing what groups are affected by any particular suggested pricing change. One of the motivations of the study on which this book reports was to begin to satisfy this interest by estimating how the use by residential households of several telephone services is associated with "demographic" characteristics of those households such as income, race, family size, and so forth. Before the editor originated this project, there was little formal evidence about how such demographics affect telephone usage; to the editor's knowledge, nothing had been published on the subject,[2] although simply by experience traffic engineers and forecasters had learned the qualitative effects of certain demographic characteristics.

A more specific motivation for this study was a proposed change in the structure of the prices of local telephone service. For all but a few cities in the U.S., the pricing of local service has been "flat rate", especially for residence customers; *i.e.*, for a fixed monthly fee, a telephone subscriber can make any number of calls of any duration within some specified geographical area. But during the early 1970's the American Telephone and Telegraph Company (AT&T) developed the policy that the operating telephone companies should plan for and implement changes in the pricing of local telephone service so that local calls have charges attached to them. This idea is often denoted "usage

---

[2] While this study was in progress, some results of two usage studies were published. One small-scale study by New York Telephone was reported in Leo Katz, "Planning Is for People", *IEEE Spectrum,* v. 12, no. 2 (February, 1975), pp. 57-60. It relates combined business and residence telephone usage in several areas of New York City to some measures of demographic composition in those areas. Preliminary results of the other usage study, by the Southern New England Telephone Company, were described in the presentation of Frederick Fagal and Robert Little to the Connecticut Public Utility Commission Authority (February 1, 1977). We should also alert the reader to a first-rate study that relates the price of telephone service and the demographics of a household not to usage but to whether a telephone is available to the household: Lewis J. Perl, "Economic and Demographic Determinants of Residential Demand for Basic Telephone Service", National Economic Research Associates, Inc. (March 28, 1978).

sensitive pricing" (USP) or "measured service". In such a plan, there are four pricing elements that are usually emphasized: (1) a charge for initiating a call; (2) a further charge related to the duration of the call; these charges also vary with (3) the distance of the call and (4) its time of day and day of week. Since USP plans are receiving a lot of attention, there is special interest, of the sort mentioned above, in what the impact of their institution will have on various groups. Thus, this study relates demographics to each of the above four USP elements.

The interest in USP not only helped to motivate the initiation of this study, but also affected its design. Chicago, the city from which the data for this study are drawn, already had a telephone pricing structure that was partially usage sensitive: there was a charge for calls within the city, although not for duration, distance, or time of day. At the beginning of this study, the Illinois Commerce Commission was considering the institution of pricing calls within Chicago according to their durations. This project was originally intended as a pilot study, preparing for the analysis of the effects of this pricing change. The anticipated before-and-after study was not only intended to reveal the impact of the pricing change on different groups, but in addition it was expected to show what the changes in traffic patterns and revenues were when the more usage sensitive rate plan was adopted. Charging by durations has not yet been instituted, so the before-and-after study has not materialized. But the existing data still provide a rich lode which is heavily mined in this book.

Still another motivation for this study resulted from the observation that there appeared to be more opposition to USP from telephone customers than seemed appropriate, given their usage. One hypothesis to explain this phenomenon is that the typical customer overestimates his telephone usage, so mistakenly calculates that USP would be harmful rather than helpful. This last suggestion is tested in the study and is found to be true.

Given the above motivations, the editor decided to proceed with the study. The overall strategy of the study had three parts: (1) Design and administer a questionnaire that would gather demographic information on a sample of individual households (during 1973). (2) Collect telephone usage data on the same set of households (1972, 1973, 1974). (3) Analyze the relationship between telephone usage and demographics (1974 to 1979).

The following section describes the present telephone rate structure in Chicago. Highlights of the empirical results occupy Section 1.3. Following that are some observations on the impact of price changes, then a discussion of the regulatory dilemma regarding whether to use information about impacts. Section 1.6 discusses benefits and limitations of the study, and Section 1.7 is a methodological overview. The chapter concludes with a brief description of the contents of the remainder of the book.

1.2 The Rate Structure of Telephone Calling for Chicago Customers

As mentioned above, at the time of the collection of this study's data, Chicago had a local rate structure that was partially usage sensitive (its rate structure today is similar). Chicago telephone customers could call anywhere in the city of Chicago for one "message unit", regardless of the duration of the call and regardless of the time of day or day of week. If they called to a specified ring of Illinois suburbs — up to roughly thirty-five miles from downtown Chicago — they incurred more than one message unit to initiate the call, then further message units related to the duration of the call. The number of initial and duration-related message units depended upon the distance of the call, but not upon the time of day or day of week. In order to compute a customer's bill for a month, the telephone company added the customer's number of local calls to his number of suburban message units; a charge for excess message units resulted — usually 5¾ cents each — if the number of message units exceeded the customer's chosen message-unit allowance. The residence customer having single-party service could choose an allowance of zero, 80, 140, or 200 message units per month, paying more the higher the allowance. The customer could choose a premium flat rate "unlimited" option. (About 5 percent of the sample subscribed to the unlimited service, paying $24.50 per month in 1974.) In addition, a two-party-line customer could choose between an allowance of 30 or 60 message units. Calls beyond the specified ring of suburbs were subject to the fully usage-sensitive toll rate schedule, with charges that varied by distance, duration, time of day, and day of week. For the purpose of this book, the calls that both originate and terminate in Chicago are referred to as "local calls", those that originate in Chicago but terminate in the specified suburban area are denoted "suburban calls", and those between Chicago and any point outside that suburban area are labeled "toll calls".

1.3 Highlights of the Results Relating Telephone Usage with Demographic Characteristics

How then does telephone usage vary across groups? Do households with teenagers use the phone more than those without teenagers? Are senior citizens glued to the phone all day? Or are there no differences in the calling habits of the groups in our sample?

In fact, we found several statistically significant differences. These are explored in depth by our colleagues and us in the remainder of the book. But here we present in a non-technical style the highlights of the results.

Subsection 1.3.1 below summarizes the results of Chapters 5 and 6, which analyze some aggregate measures of individual households' local and suburban telephone usage. Subsequent subsections deal with local and suburban calling patterns by time of day and by distance, with bills, and with toll usage.

1.3.1 Aggregate Local and Aggregate Suburban Usage

Several demographic characteristics were found to be significantly related to telephone usage. The characteristics that we have selected to be discussed in detail in this subsection are race, family income, age of the head of household, and household composition (meaning not only the number of persons in the household, but also the age and sex of household members). But other characteristics are also estimated to be significantly associated with telephone usage, including certain occupations, whether one has lived in Chicago less than a year, and employment status. The results for the first list of characteristics are these:

(1) Local telephone usage is greater in many dimensions for black households than for white households: the median number of local calls is almost twice as large; the median of the average durations of local calls is about one quarter larger; and, as one might expect from the above two comparisons, the median total local conversation time is larger by more than a factor of two. One might speculate that the differences in usage patterns between the races are due to other demographic characteristics such as income, family size, *etc.;* yet multiple regression analyses that control for a wide variety of other demographic characteristics still reveal race to have a large influence.

While the local usage of black customers exceeds that of white customers, calling to the suburbs by white customers is dramatically higher than that by black customers. In terms of medians, the former make over four times as many of such calls as the latter, talk over twice as long per call, have a total suburban conversation time that is over eighteen times greater, and have a total number of suburban message units that is over five times larger. If one looks at total message units — the sum of the number of local calls and suburban message units — one finds a statistically insignificant difference between the medians for the two racial groups. Multiple regressions on suburban message units and on total message units find similar results, even though other characteristics are held constant. The conjecture is that much of the difference between the races in local versus suburban calling is explainable in terms of their communities of interest; the suburbs tend to have a higher proportion of households that are white and high-income than the city of Chicago does.

(2) As income rises, the median number of local calls tends to rise, although no pair of income groups is significantly different. A high income is also associated with low average durations. The results from multiple regressions are different: controlling for other variables, the number of local calls is estimated to be significantly higher only for incomes of over $30,000; total conversation time is higher for the $15,000-to-$19,999 group; and there is no significant income effect for the average duration of local calls.

Income has a strong positive association with the level of suburban calling: the medians of the number of suburban calls, of total suburban conversation time, and of suburban message units all rise sharply with income. The median of total local and suburban message units also rises with income. From multiple regressions, the results are that both suburban message units and total message units are higher for households with an income of over $30,000 than for others.

(3) Some interesting results have also been found for the age of the head of household. The medians of the number of local calls and of total local conversation time are lower for the two age groups 55 to 64 and 65 or over as compared to the younger age groups. And the medians of total local and suburban message units display a strong downward trend with increasing age.

(4) Multiple regression analysis indicates that local usage is strongly related to the number of persons in a household, other things equal, and that it makes a difference what the ages and sexes of the household members are. The group with the largest effect is young teenage girls. Further, total message units are particularly high where there are young teenagers of either sex.

(5) Although demographic variables were isolated that are correlated with telephone usage, it is difficult for these models, containing a few demographic variables, to predict an individual customer's

telephone usage with any large degree of precision. A related point is that the variation in the telephone usage among households within a demographic group is large relative to the differences between groups. Still, such models may provide estimates for entire groups of customers with good accuracy.

(6) Most sampled customers estimate an average number of local calls they make per day that is larger than the number actually recorded for them by the telephone company. Most sample customers also over-estimate the average duration of their local calls.

1.3.2 Local and Suburban Usage by Time of Day

For brevity, here we discuss results for only two measures of calling: the mean of customers' total local conversation times summarizes local usage, and the mean of suburban message units summarizes suburban usage.

(1) Mean total local conversation time for weekdays for the entire sample displays a broad peak of 2.0 minutes per hour, for both 7-8 p.m. and 8-9 p.m. For most of the rest of the day, usage is a little over one minute per hour, although during the interval of 4 to 7 p.m. it climbs to its peak. Usage between 11 p.m. and 8 a.m. is very low.

(2) The hour of the peak in total local conversation time is the same for the two racial groups, but the black group has much greater usage in the evening relative to daytime usage than the white group does; the level of their evening peak is also twice that of the white group.

(3) The quantitative pattern across hours is similar among the income groups. However, the under-$5,000 group has the highest usage for most of the day-time hours.

(4) Households headed by persons of age 65 or over have a peak of total local conversation time in the evening, whereas they have a morning peak for the number of local calls. Still, the difference between morning and evening total local conversation time is less pronounced for them than it is for households with younger heads.

(5) The peak in suburban message units for the entire sample during weekdays appears to be at 7-8 p.m.

(6) For most hours of the day, white customers have substantially greater suburban message unit usage than do blacks, and the differential is greater in the evening than during the day.

1.3.3 Local and Suburban Usage by Distance

(1) Non-toll calls are heavily concentrated near the customer's location. For example, on the average, 31 percent of total local and suburban conversation time terminates within a customer's own Central Office Area, and 62 percent terminate in Central Offices within five miles of a customer's originating Office. As distance of a mileage band from the originating Central Office increases, the percentage of calls terminating in that mileage band decreases.

(2) The average duration of calls rises up to the 5-10-mile band, but then declines as mileage increases beyond 10 miles.

(3) Demographic characteristics tend to have only mild effects on the distance of calls. For instance, no significant differences are found in the average distances of calls between any of the studied demographic groups.

1.3.4 Total Charges and Vertical Service Charges

The relationship between bills and demographic characteristics is examined by multivariate regression.

(1) Holding other things constant, total charges continually (a) increase with income, (b) decrease with the age of the head of household, and (c) increase with the number of household members who are at least 10 years of age, if the race of the household is not black; total charges also (d) are higher for black households than for white households, and (e) are higher for households whose heads have lived in Chicago for fewer than 15 years than for the others, so long as the household has four or fewer members at least 10 years of age.

(2) Charges for extras such as extensions, Touch-Tone®, and Trimline® telephones tend to be higher for households that (a) have a high income, (b) are black, (c) have heads who have lived in Chicago over 15 years, if they are under 45 years of age, (d) have two members rather than one member at least 10 years of age, (e) have heads of age 25 to 34 rather than over 44, or of age 35 to 44 rather than over 64, if they have lived in Chicago over 15 years, (f) have heads of age under 25 rather than over, if they have lived in Chicago 1 to 15 years, (g) have heads who are separated rather than either single or married, or who are married rather than single.

1.3.5 Toll Usage

(1) The probability that a household made at least one toll call in the month of study is lower if the household is white rather than black and if the household has lived in Chicago a long time rather than a short time.

(2) Considering the set of customers in the sample who made at least one toll call, the medians of three measures of toll usage — the number of toll calls, total toll conversation time, and total toll charges — are all significantly higher for households with annual incomes of $20,000 or more than for those with lower incomes.

1.3.6 Remarks on the Empirical Results

There are tremendous differences in telephone calling patterns between individual households, even between those that are identical in all the respects that the study has measured. But many of the estimated relationships of telephone usage and bills with demographic characteristics are statistically significant and strong. In fact, many estimated differences between groups are on the order of a factor of two or more. Thus there is strong evidence that demographic characteristics can be used to explain or predict at least aggregate calling patterns of households. Questions remain as to whether demographic characteristics have a causal effect themselves, or are rather proxies for some other causal effects. For example, it is possible that some of the age effect that is estimated really reflects calling behaviors that were developed in youth; it might be the case that people who are old now developed their habits long ago when telephones were little used, while people who are young now developed their habits recently when telephones are heavily used. This is not to say that the kind of estimates in this book are not useful. The point is that their usefulness is in direct proportion to the degree of reliability as proxies of such easily obtained data as age.

1.4 Observations on the Impact of Pricing Changes

In the first section we said that regulators and telephone companies are interested in knowing the impact that a price change has on various groups. An evaluation of the estimated relationships of various measures of telephone usage with demographics can give a reasonable impression of the impact that some simple price changes are likely to have on the studied groups. So the reader can make reasonable inferences of qualitative impacts of many simple price changes by referring to the figures in Chapters 5 through 10. But analyzing complicated price changes is liable to be safer if one uses the raw data.

One should keep in mind that if any price were to change, then a customer's calling behavior is likely to change. This study provides no estimates of the degree to which calling is affected by price changes, so we cannot estimate changes in bills.[3] According to economic theory, a change in bill is a poor measure of the impact on a customer's welfare anyway. As a customer reduces his calling after a rate increase, he reduces his expenditure relative to what it would have been if he had maintained his demand. But as he does so, he also forgoes the benefits he was receiving from the now-discontinued usage. Thus, the customer's change in bill understates the reduction in his welfare due to the rate increase.

To many economists, a more reasonable measure of impact is the change in a consumer's surplus. The aggregate or "consumers'" surplus that is derived from a good is defined as the amount that consumers would be willing to pay to consume the quantity of the good they do rather than go without it altogether, minus what they do pay. For small price changes, the impact on a consumer's surplus due to a change in price is very close to what the increase in the telephone bill of the consumer would be if his usage and chosen message unit allowance remained exactly the same after as before the change in prices. Such a calculation is called a "repricing" in traditional telephone rate making. For illustration, we performed a repricing experiment using the present sample of customers. We estimated the repricing impact across three income groups: under $5,000, $5,000 to $14,999, and $15,000 or over. For three illustrative usage sensitive price changes, the qualitative results are these: (1) Suppose that the price structure for calls within Chicago were changed by having the message unit charge rise with the distance of each call. Taking into account where each customer's message unit usage is before the change relative to his chosen allowance, the average increase in bill is virtually identical for all three income groups in the sample. (2) Alternatively, suppose that a message unit premium were applied to both local and suburban calls made between 8 a.m. and 10 p.m. on weekdays, and that an equal discount were applied to calls made during all other hours. In this instance, the average increase in the repriced bill is larger for those

---

[3] See Chapter 11 for a proposed technique for obtaining such an estimate from the kind of data used in this study.

with larger incomes; in fact, the estimated increase in bill for the group with larger incomes of $15,000 or more is twice as large as the increase for those with incomes of less than $5,000. (3) A very different outcome is estimated if it is supposed that an additional message unit charge is introduced for each 5 minutes in duration of a call within Chicago. The estimated increase in the repriced bill is smaller for households with higher incomes. Thus, from these examples, it is clear that the relative reprice impact on the income groups can be considerably affected by the nature of the suggested change in price.

1.5 The Regulatory Dilemma

The above three examples of estimating the impact of pricing changes on different demographic groups are merely illustrative. Demographic groupings other than by income may be of interest. Further, the four common pricing elements — a charge for initiating a call, and charges by duration, distance, and time of day — give the rate maker or regulator a large menu of price structures to consider. We hope that this study will give quantitative input when choices from the menu are weighed against one another.

For any pricing change, one may view the reprice impact that is calculated for each customer as the sum of two components. (1) Suppose that one computes the increase (or decrease) in reprice revenue as a result of the pricing change. Define the "revenue effect" as the income that is imagined to be given up (or received) equally by each customer so as to produce the reprice revenue. (2) The "redistributive effect" is the transfer of income among customers that causes the sum of the revenue and redistributive effects for each customer to equal the reprice impact that was calculated for him.

Instituting changes in price structure presents the policy maker with a classic dilemma. On the one hand, a move toward more cost-related or "allocatively efficient" prices is liable to trigger off redistributive effects that may be unanticipated or may be considered to be "unfair". On the other hand, a purposeful move toward "fairer" rates, or the avoidance of a move toward more efficient rates so as not to generate opposition from particular groups, may result in a waste of economic resources.[4] For example, the policy maker may be reluctant to institute charges by the number of calls in "flat rate" areas because of the objections of those whose telephone bills would be increased.

---

[4]For an overview of this issue, see, *e.g.*, Edward E. Zajac, *Fairness or Efficiency: An Introduction to Public Utility Pricing* (Cambridge: Ballinger, 1978).

But then a certain number of calls will continue to be made even though the cost to the local telephone network of carrying those calls exceeds the value of the calls to the parties involved. Since revenues must cover costs, so long as the cost of metering is not very high the average subscriber would be better off if these low-value calls were eliminated by attaching a price to them.

The dilemma that confronts the regulatory policy maker is similar to that of the tax designer, and is exacerbated by the difficulty of calculating economically efficient rates or taxes. The theory of such calculations as well as the theory of group action in both taxation and regulation is the subject of an extensive literature.[5] We shall not attempt to discuss this literature here, nor shall we try to resolve the regulatory dilemma as it occurs in the pricing of residential telephone service. However, the following two observations are important and should be kept in mind in any applications of our results:

(1) Our data indicate that telephone prices are generally an ineffective means of redistributing income among groups. The reason is that the variation of usage within each demographic group is typically large relative to the variation between groups. Thus, by some pricing action, on the average one might succeed in benefiting the customers in one group and adversely affecting those in another; but there will be a large number of customers in the "deserving" group — in some instances even a majority — who are hurt rather than helped, and there will be a large number in the "undeserving" group who are helped. Illustrative of this point is the fact that many of the customers in California who are subscribing to the low-cost Lifeline service are high-income professional persons.[6]

(2) The usage patterns that we have observed may change with changes in price structure and level. We feel that this point should be emphasized here, especially with the advent of more competition in telecommunications. This new reality will undoubtedly limit the policy maker's ability to set prices higher than costs on some services to enable prices to be low on others.

---

[5] See, for example, Stigler, *op. cit.*; v. 6 (1976) of the *Journal of Public Economics;* and Bruce M. Owen and Ronald Braeutigam, *The Regulation Game* (Cambridge, Mass.: Ballinger, 1978).

[6] Pacific Telephone and Telegraph Co., *Lifeline 1976: Characteristics of Residence Subscribers,* General Administration Accounting, Market Research and Statistics, Project 6-12 (June, 1976).

1.6 Benefits and Limitations of the Study

Two of the benefits that may derive from the study have already been described in the above discussion of the motivation for the project in Section 1.1 above: (1) The results may begin to satisfy a desire on the part of regulators and telephone companies to know to what extent different demographic groups are affected by rate changes, especially toward usage sensitive pricing; and (2) the study has verified the hypothesis that telephone customers tend to overestimate their local telephone usage. Because the before-and-after study of the effects of charging by the duration of local calls did not materialize, the study does not provide any direct estimates of what the effect on usage would be due to extending usage sensitive pricing. But several other benefits, while not originally motivating the study, are important spin-offs: (1) It may be possible to predict the telephone usage patterns in other cities whose demographic compositions differ from Chicago's if the telephone companies serving those other cities were to adopt the rate structure in the studied area; these predicted usage patterns might be disaggregated by time of day or, less confidently, by distance. In turn, such predictions might be useful for aiding telephone pricing decisions and traffic engineering. Whether such predictions would be safe is, with our present knowledge, speculation. There is some evidence discussed in Section 11.2 that prediction with demographics alone might be risky. (2) Demographic groups that heavily use certain telephone services may be identified for marketing purposes. (3) Usage may be estimated inexpensively at the neighborhood level to aid traffic engineering. (4) Important demographic characteristics associated with the demand for telephone service are identified, so that similar studies in the future might be carried out at lower cost in two ways. Firstly, the number of questions in the questionnaire could be reduced. Secondly, a given accuracy of prediction can be achieved with a smaller number of customers in the sample by sampling the customers in the various Central Office Areas with different probabilities, depending upon the demographic composition in each Area.

In order to present a balanced view, it is also useful to list some limitations of this study. One difficulty may be statistical bias. It is not known that there is bias, but there are at least four potential sources: Firstly, the sample is restricted to the set of customers who are served by particular types of switching machines. Secondly, about a quarter of the sample did not respond to the questionnaire. Thirdly, a fifth of the sample had its telephones disconnected during the three-year interval of the study. Fourthly, while not presenting any difficulties for uses within Chicago, there may be special features of Chicago that influence telephone usage but are not reproduced elsewhere. Related to that point is the limitation that the work to date provides estimates which apply only to areas with Chicago's rate structure and rate level. One more difficulty that may be worth mentioning is that any extrapolation from the sample requires either the administration of a questionnaire, which is somewhat expensive, or the reliance on Census data, which may be out of date and which pertain to a larger population than those with telephones.

1.7 A Methodological Overview

In this section we discuss some methodological issues that are common to all of the analytical chapters (numbers 5 through 10). We also briefly survey the various analytical techniques that are employed in this book.

One general methodological issue is whether to relate each telephone measure to one demographic characteristic at a time or to several simultaneously. Both kinds of analyses are performed in this book to address two types of questions that are often asked, although it is found that the qualitative results from the two kinds of analyses seldom differ. On the one hand, one may be interested in comparing the calling behavior of two divisions of a single demographic characteristic, say the number of calls made by households with an income of under $5,000 versus those made by households with an income of $5,000 to $9,000. This kind of comparison is generally what is desired when one investigates the impact of pricing changes among demographic groups. Single characteristic analysis is also easily understood by the non-technical person.

On the other hand, if one simply compares the calling rate among income groups, then one reasonably suspects that the differences that one observes are at least partly due to some other demographic characteristic. For instance, it might be the case that a lower calling rate by low-income households than by high-income households might really be due to senior citizens' tending to have both a low calling rate and a low income. To summarize calling patterns of several demographic characteristics at once, one uses multiple regression analysis. This methodology allows one to "control" for all but one factor, so that one can make statements such as: "Given households that are alike in all other respects, those with higher incomes make more calls." Multiple-characteristic analysis thus moves one closer to the basic forces that influence telephone usage than does single-characteristic analysis. It also is more useful for drawing inferences beyond the sample if the sample does not have the same composition as a larger population.

Now let us turn to another issue — commonly called the "multiple comparisons" problem. As an illustration, consider the toss of two dice, say Die A and Die B. There is one chance in 36 that both will come up a five. If Die C is added and the three dice are tossed, there is still one chance in 36 that both A and B will come up a five, but also one in 36 of A and C both a five, and one in 36 of B and C both a five. Thus, when three dice are tossed, the probability that *any* pair is both a five is greater than the one in 36 that an *earmarked* pair is both a five; in fact it is $3 \times 1/36$ or one in twelve.

This simple example illustrates that a comparison of two earmarked random events can have very different statistics from a comparison of any two of several random events. In interpreting the statistical results that are reported, this multiple comparison issue should be kept clearly in mind. Suppose for example that one is considering either more than one pair of demographic groups or more than one telephone usage measure. Whether there is a multiple comparisons problem depends upon what question one wants answered. On the one hand, there is no conceptually different problem from that in the single comparison case if one asks the question, "What is the set of comparisons for which there are significant differences?" On the other hand, a difficulty does arise if one asks a question similar to the following: "Is there *at least one* comparison for which a statistically significant difference is found?" Three examples of this kind of question are:

(1) Is there any measure that is significantly affected by this demographic characteristic?

(2) Is there any pair of divisions of this demographic characteristic that is significant, so I can conclude that the characteristic is important?

(3) Is there any hour of the day for which there is a significant difference between these two groups?

For questions like these, the statistical tests applicable to a single, earmarked comparison do not apply. When the underlying populations are not different from each other, if a 5 percent significance level is the standard applied to each individual comparison, then the answer to the above kind of question will turn out to be affirmative with a probability of greater than 5 percent. For instance, in the case of two independent comparisons, the probability of finding at least one "significant" difference among the two sample comparisons when there are no differences in their respective populations is not 5 percent but close to 10 percent:

$$Pr\ (at\ least\ one\ is\ significant) = 1 - Pr\ (neither\ is\ significant)$$
$$= 1 - (0.95)^2$$
$$= 0.0975.$$

In this example, if one wants to apply a 5 percent standard to the multiple comparisons, one should use the much tighter 2.5 percent standard (approximately) for the two individual comparisons. Extending this example, the standard for individual comparisons should be tighter the larger is the number of independent comparisons being made.

The multiple comparisons problem is less serious when the outcome of each of the multiple comparisons is independent of all the others. (An "outcome" is a finding of a significant or insignificant difference, regardless of the direction of the difference.) There are three mechanisms that cause a positive correlation between the outcomes of the comparisons in this book. Firstly, for a given measure, outcomes of comparisons for various demographic characteristics are positively correlated because the same customers enter all comparisons. Secondly, the outcome of a comparison for one measure is positively correlated with that for most of the others because one measure of a customer's usage is usually correlated (positively or negatively) with that of another. Thirdly, the outcome of the comparison between two divisions of a demographic characteristic is positively correlated with the outcome of the comparison between one of these divisions and another: whether the usage measure for the division that is common to the comparisons randomly turns out to be high or low affects the outcome of both comparisons.

The dependence of each comparison with others reduces the probability that at least one comparison produces a finding of a significant difference when there is no difference between the underlying populations. The multiple comparison problem still exists, but it is quantitatively not so serious as the assumption of independence implies.

For some demographic characteristics with more than two divisions, there is another point that should be borne in mind when evaluating the statistical significance of differences between divisions. The divisions of some characteristics have a natural ordering; examples are income, age, and the number of persons in a household. For such characteristics, there is more information than is contained simply in pairwise comparisons of their divisions. For instance, if one observes for some sample that the median calling rate rises steadily with income, then one would tend to attach more credence to the estimated pattern than is indicated by pairwise significance tests. But if the medians displayed no pattern, then one would tend to attach less credence to any observed "significant" difference.

A more technical issue which arises for Chapters 5, 6, 7, 8, and 10 is that the recorded durations for local, suburban, and toll calls tend to exceed the actual durations. The reason is that, before the data were sent to us, the duration of each call was rounded to an integer by a convention that is appropriate for billing purposes. Depending upon what question one is addressing, one may or may not want to correct for this rounding process. The authors of Chapter 10 chose not to correct the bias; the authors of Chapters 5 through 8 did attempt to eliminate it. The procedure that is used in Chapters 5, 7, and 8 is described in detail in Section 5.4.2. That for Chapter 6 is described in Section 6.2.

Now we turn to an outline of the analytical techniques that are used in the book. This section provides only a non-technical overview. For each technique, we refer to the first chapter in which the technique is used, where it is discussed in detail. We should mention that all of the techniques are to be found in some form in the statistics or the econometrics literature; however, some are not very well known, and others are extensions of known techniques.

The most heavily used technique in this book is the "box plot". It graphically displays a five-point summary of the distributional character of a set of data on some variable, with special emphasis on the median value (for instance, the median number of calls for a set of households would be the number of calls made by the household for which the number of households with higher usage equals the number with lower usage). The most frequent use of the box plot here is to compare the telephone usage among the divisions of a demographic characteristic. "Notches" are added to the box plots to provide a rough guide to whether a difference that one observes is statistically significant. A detailed explanation of the box plot can be found in Section 5.3. A graphical indication of sample size is explained in Section 10.3.2. In some chapters, the box plots are supplemented by a non-parametric test for differences in medians — the Wilcoxon-Mann-Whitney Rank Sum Test — which is explained in Appendix 5.A.

A technique that is used in Chapters 3, 7, and 8 is the comparison of means. Many of these comparisons are accompanied by the well-known Normal test for a difference in means. In Chapter 7, however, comparisons of calling patterns by hour of the day are presented with error bars as rough visual guides to statistical significance. The latter procedure is described in detail in Section 7.4.

Section 6.3 explains the multiple regression analysis of Chapter 6, which models the error terms as having a non-constant variance and as being correlated across months. Chapter 9 also applies multiple regression analysis, along with "Tobit" analysis, which attempts to account for the set of households that have no charges of a particular kind. Explanations are in Section 9.3.

Chapter 10 compares the demographic mix of the set of households that made at least one toll call to those who made none (the analysis is repeated for suburban calls). A formal test accompanies the comparisons (see Section 10.3.1).

1.8 Preview

In the following chapter, the sampling plan for this study is outlined. Chapter 3 describes the questionnaire design and administration, as well as the collection of calling and billing data. The data base and its management are presented in Chapter 4. Chapters 5 through 10 contain the empirical results of the study. They all relate household demographic characteristics to measures of the use of telephone services. Chapters 5 and 6 examine aggregate local and aggregate suburban usage, one by comparisons of medians and the other by multiple regressions. Chapters 7 and 8 go on to display local and suburban usage by time of day and by distance, respectively. Chapter 9 presents multiple regression analyses of total bills and vertical service charges. Toll usage is the subject of Chapter 10. The last chapter discusses other studies, suggests improvements that might be made in similar studies in the future, and suggests extensions of the present analyses.

CHAPTER 2

The Development of the Customer Sample

Wm. H. Williams and Michael L. Goodman

2.1 Introduction

For the study on which this book reports, a special sample of 849 residence telephone lines was drawn in 1972 and 1973. The draw was from the set of Chicago telephone lines that were served by certain types of switching machines. These machines served only a part of the city.

This chapter describes the sampling procedure and explains the background of the decision to restrict the area of the sample. Following that is a presentation of some demography of the sample population as compared to the entire city of Chicago. At the end is a note by the editor regarding inferences beyond the sample.

2.2 Relevant Characteristics of Telephone Equipment in the City of Chicago

At the time of the study, the city of Chicago was served by 35 Central Offices. A Central Office Area can be thought of as the area in which customers are served by a central building that houses switching equipment. These buildings are sometimes called "wire centers". The 35 Central Office Areas divide the area of Chicago into 35 (approximately) mutually exclusive and exhaustive geographical regions. Each of the Central Office Areas contains several prefixes (the first three digits of the telephone number). Of the 35 Offices, six serve very few residence customers, and, since the focus of this study was on residence usage and not business, these six Offices were excluded from the study. Twenty-nine Offices remain.

There are three major types of switching equipment in Chicago, #1 Crossbar, #5 Crossbar, and #1 ESS. These switching machines differ in ways that are important to studies that require calling detail. In particular, it is critical that the time, duration, and destination of local calls made by the sample customer be observed. Such observations are easy to make with #5 Crossbar and #1 ESS switching equipment but at the time of the study were very difficult to make in #1 Crossbar Offices, which in 1972 served about 94 percent of Chicago residence customers. The #5 Crossbar and #1 ESS machines will be referred to as the "easy" equipment; #1 Crossbar will be called the "difficult" equipment.

2.3 The Sample

2.3.1 Easy Equipment

The sample was selected from the customers served by easy equipment. No sample customers were selected who were served by #1 Crossbar equipment. There were various factors involved in this decision. On the one hand, it was technically very straightforward to make the requisite observations from the easy equipment; it would have been quite expensive and time consuming to develop measuring equipment for #1 Crossbar switching equipment. In addition, since this was originally designed as a pilot study, it was felt that the study of various demographic subgroups was a more important goal than obtaining a representative sample of the city of Chicago. Finally, if it had turned out to be necessary to simulate the distribution of the demography of Chicago as a whole, it is theoretically possible to do this by appropriate reweighting of the available observations. On the other hand, sample reweighting requires demographic variables with *known* distributions. This knowledge permits the appropriate comparisons and subsequent reweighting. Without some reweighting, it is entirely possible that the sample would be quite unrepresentative on important variables. This is an argument for drawing a sample from the entire city of Chicago.

In spite of these latter arguments the sample was drawn from the easy equipment only.

2.3.2 The Sample Draw

The sample of customers was drawn systematically from an October, 1972, #5 Crossbar and #1 ESS printout of customers supplied by the Illinois Bell Telephone Company, where the customers in the printout were in order of their telephone numbers.[1] This draw resulted in a sample of 522 business separately-billed telephone numbers and about 4,000 residence lines. All of the business numbers were retained for future use (to date no analysis has been performed with them). The sample of 4,000 residence lines was much too large for current purposes. If post-stratified by prefix, these 4,000 residence lines would be distributed over the prefixes in approximate proportion to the number of residence lines appearing on the printout. Since we wanted the sample to be proportional to the total number of lines in each prefix, and since a sample of 4,000 was too large, a stratified random sample was drawn from the 4,000 in such a way that the size of the sample from each prefix was proportional to the total number of residence lines in the prefix. Each within-prefix sample was selected by listing the 4,000 lines by prefix and using simple random number generation. The resulting size of the October, 1972, residence sample was 717.

For analysis it seems unlikely that the systematic order in which the units were selected from the printouts is correlated with any of the variables of interest; consequently the sample data were analyzed as a stratified random sample. Furthermore, since the sample is proportionally allocated among prefixes, efficient and unbiased inferences for the sampled prefixes as a whole are obtained simply by calculating unweighted sample statistics.

In April, 1973, 132 more sample customers were randomly drawn from two additional Central Office Areas. The proportion of customers that were sampled was the same as for the original Areas. The combined sample size was then 849. This supplementary sample is discussed more fully in Chapter 3.

2.4 Some Demography of Chicago

It would be of some interest to know how the demographic characteristics of the telephone customers in the sample compare to those of the customers in the rest of Chicago. Lacking such data on

---

[1] The draw consisted of the tenth, twentieth, and thirtieth telephone numbers from each page.

telephone customers for the rest of Chicago, one can examine data for the geographical areas from which the sample was and was not drawn.

There were two major sources of demographic data, the U. S. Bureau of the Census and the Illinois Bell Telephone Company. Information on total population, race, housing rental, value of homes, and plumbing was obtained from the Census Bureau. The number of residence main telephones in service for each of the 29 Central Office Areas was obtained from the Illinois Bell Telephone Company.

An operational difficulty was that the Census Bureau reports data by city blocks and not by the geographical regions defined by the 29 Central Office Areas. These city blocks can be combined into block groups and then into census tracts and finally into neighborhoods. There are approximately 900 census tracts in Chicago and exactly 76 neighborhoods.

The boundaries of the 29 Central Office Areas are independent of the census and often cut across census boundaries. However, for each block of the census data, it was possible to break down the census data to correspond to telephone Central Office Areas. The transformation used to convert the data on the 76 neighborhoods into data on 29 Central Office Areas is described below.

To illustrate, suppose that neighborhood X has 10,000 residential housing units of which 9,000 are in Central Office Area A and 1,000 are in Area B. Similarly, suppose that neighborhood Y has 2,000 in A and 3,000 in B. This breakdown is shown below.

		Neighborhood	
		X	Y
Central Office Area	A	9000	2000
	B	1000	3000
Total		10,000	5000

Assuming that Central Office Area A covers only neighborhoods X and Y, then the data for neighborhoods is converted into data for Central Office Areas by the equation

$$N_A = .9N_X + .4N_Y,$$

where N is the number of persons with the characteristic of interest, and .9 and .4 are the respective fractions of Neighborhoods X and Y that overlap with Central Office Area A. So for example, if there are

8,000 white households living in neighborhood X and 12,000 white households living in neighborhood Y, then we estimate that there are $12,000 = .9 \times 8,000 + .4 \times 12,000$ white households living in the area defined by Central Office Area A. Exhibit 2.1 includes some Central Office Area demographic estimates derived from neighborhood data in this way. We developed estimates (not shown) in the same manner for the distributions of monthly rents and of housing values.

Exhibit 2.2 displays some selected estimated characteristics of the area as a whole served by the easy equipment, and similar estimates for the difficult equipment and for Chicago as a whole. Appendix 2.A explains the calculation of the estimates, including the necessary allocation of population between the easy and difficult equipment. (The Beverly and Pullman Central Office Areas, into which ESS machines were introduced in 1973, are counted as 17.9 percent and 18.4 percent ESS, respectively.) Consideration of Exhibit 2.2 shows that the area served by the easy machines (#5 Crossbar and #1 ESS) have a population that is 40 percent non-white while the area served by #1 Crossbar Offices is 34 percent non-white. In addition, the easy area has an overabundance of very high and very low rents. Thus, the area covered by #5 Crossbar and #1 ESS switching equipment has demographic characteristics which are somewhat different from the area served by #1 Crossbar equipment.

Exhibit 2.1

Chicago Central Office Area Demographic Information

Name of Central Office Area	(1) Res. Main Stations Total	(2) #5XB #1 ESS	(3) Number of Households	(4) Population	(5) Percent Non-White Persons	(6) Percent No Private Toilet
Beverly	37,771	*	35,427	122,718	37.90	.26
Calumet	17,009	16,991	18,765	53,488	59.40	4.37
Franklin	53	8	579	1,759	61.74	17.98
Hyde Park	56,563	1,332	72,080	177,005	78.32	4.85
Il.-Dearborn	3,443	3,434	3,428	4,427	38.70	9.28
Kedzie	23,142	794	35,820	121,262	86.66	4.53
Lawndale	25,132	334	35,242	123,463	69.53	1.79
Merrimac	30,873	2,504	31,703	88,375	10.62	1.09
Monroe	27,903	9,919	43,955	138,549	44.79	7.65
Portsmouth	34,852	8,947	32,050	106,083	2.29	.46
Prospect	54,046	2,187	59,448	180,450	22.19	1.03
Pullman	38,391	*	38,253	126,292	47.48	1.02
Superior	30,236	11,599	43,342	79,826	34.00	8.43
Wabash	1,496	1,132	2,744	6,684	36.82	27.11
Austin	31,307		37,517	108,934	38.74	2.43
Edgewater	54,713		71,081	153,696	6.93	4.70
Humboldt	67,377		94,373	268,556	7.97	3.26
Irving	40,931		43,592	117,878	1.91	.78
Kildare	56,251		58,729	163,071	.66	.66
Lafayette	37,739		46,409	134,576	4.04	1.97
Lakeview	81,703		103,566	217,526	7.14	3.85
Ludlow 6	7,327		6,926	23,584	5.16	.75
Mitchell	3,816		3,530	11,764	.43	1.71
National 5	6,286		6,549	22,444	1.04	.17
Newcastle	28,055		25,799	77,869	.30	.17
Oakland	46,751		66,583	194,333	84.35	7.47
Rogers Park	58,745		58,284	136,265	3.70	1.57
So. Chicago	53,865		56,869	167,051	37.78	1.80
Stewart	71,085		74,908	238,406	88.83	1.31

Sources: (1) "Chicago Operations Central Office In-Service Information" July, 1972, IBT Form CFD 370S (1-72)
        (2) Planning and Rate Administration, IBT
        (3) "1970 Census Control Sheets", IBT
    (4)(5)(6) Transformation of neighborhood data into Central Office data. Neighborhood data taken from "1970 Census, Chicago Community Areas Only, 1st Count Summary Tapes", Research and Statistics Division, Chicago Association of Commerce and Industry.

* During 1973, part of the Beverly and Pullman Central Office Areas were converted to #1 ESS.

Exhibit 2.2

Comparison of Central Office Area Demography in Chicago

	#5XB + #1 ESS Areas	#1XB Areas	City Total
% Non-White Persons	40.24	33.79	34.25
% No Private Toilet	3.37	2.91	2.96
Distribution of Monthly Rents			
Under $100	52.47	41.63	42.44
$100-$149	24.69	42.46	41.13
$150-$199	11.23	11.49	11.47
$200-$299	7.39	3.32	3.62
Over $340	4.22	1.10	1.34
Distribution of Housing Values			
Under $15,000	14.87	15.10	15.09
$15,000-$24,999	60.00	54.52	54.91
$25,000-$34,999	19.99	23.02	22.80
$35,000-$49,999	4.14	6.04	5.90
Over $50,000	1.00	1.32	1.30

Source: Same as Columns (4), (5), (6) of Exhibit 2.1

Editor's Note Regarding Inferences Beyond the Sample

Consider the following question: Given the restricted sample area, are the estimates in this book capable of providing unbiased inferences and predictions beyond the sample? The answer to the question in one dimension is a qualified No: none of the estimates directly yield inferences or predictions beyond the dates of the calling data, although one may obtain them if one has some information outside these sample data (*e.g.*, the usage of all groups grows at 3 percent per year). In our subsequent discussion of the issue of inferences beyond the sample, we shall ignore sources of bias other than the restricted sample area.

Going beyond the set of customers in the sample is more interesting. The simplest case to consider is whether the estimates provide unbiased inferences and predictions in the study period for the population of customers who are served by easy machines. As stated in Section 2.3.2, because we have a proportional random sample, the answer is yes. The more difficult question is whether we can make inferences and predictions outside that population, say for Chicago as a whole.

One part of the answer is clear: Since the demographic characteristics of the easy area and of the difficult area differ somewhat, unweighted sample statistics for some telephone measure, aggregating over all groups in our sample, would differ systematically from the statistics for that measure from Chicago as a whole, even if only somewhat.

Are estimated differences between demographic groups unbiased inferences for Chicago? The condition that must be satisfied for that to be true for some analysis is that either any variables which affect usage and which are left out of the analysis are uncorrelated with the included variables, or that, for given values of the set of included variables, the left-out variables have the same expected values in the sample area as in the rest of Chicago. Intuitively, multiple-characteristic analysis seems likely to satisfy this condition more closely than single-characteristic analysis because the former leaves out fewer variables. It seems plausible that such an analysis can produce a good inference for the rest of Chicago; we have neither thought of nor heard of any strong arguments as to why such inferences should be biased. Still, there may be some systematic effect of which we are unaware.

Now consider a slight extension of the previous question. If, for the purpose of discussion, we have unbiased inferences of differences in usage between groups, do we also have unbiased inferences of the *level* of usage for each group? In order to answer yes, we must have an additional condition satisfied: the left-out variables that affect usage, even though *uncorrelated* with the included variables, should have the same expected values inside the sample area as outside, given the values of the included variables. Again, we are unaware of strong violations of this condition but should not assert its impossibility.

So far the discussion has been in regard to Chicago. A natural extension of this line of thinking is to consider the possibility of inference or prediction outside of Chicago. The models in the book are presumably inadequate for predicting the usage in other cities with rate structures that differ substantially from Chicago's because a customer's usage should be affected by rates. But a conjecture that is left as an open question for future research is that the present model can be used to predict what usage would be in other cities if they were to adopt Chicago's rate structure. There is some evidence discussed in Chapter 11 that a prediction for another city using demographics alone may be inadequate.

Appendix 2.A

The Calculations for Exhibit 2.2

Section 2.4 presents the results of calculations of demographic characteristics of the areas served by easy equipment and by difficult equipment. Here we give the mathematics of those calculations.

What prevents the calculations from being trivial is the necessity to allocate the population in the Central Office Areas between the easy and difficult equipment. Since there is no Central Office Area that is served exclusively by easy equipment, this allocation is done for every Area that has some easy equipment. At the times of the sample draw — October, 1972, and April, 1973 — the practice in Chicago was to assign the telephone lines in a Central Office Area to the various switching machines randomly. (Random assignment may not be followed now that the Custom Calling service, requiring ESS machines, has been introduced in Chicago.) Given that practice, the population in each Central Office Area should be allocated to easy and difficult equipment in the same proportions in which residence lines in the Area are served by those types of equipment. Thus, the fraction of persons or relevant households with some characteristic c in the area served by easy equipment is estimated by

$$f_c = \frac{\sum_{i=1}^{N} n_{ic} \dfrac{m_{ie}}{m_i}}{\sum_{i=1}^{N} n_i},$$

where N ≡ the number of Central Office Areas,

n_{ic} ≡ the number of persons or relevant households with characteristic c in Area i,

n_i ≡ the total number of persons or relevant households in Area i,

m_{ie} ≡ the number of residence telephone lines served by easy equipment in Area i, and

m_i ≡ the total number of residence telephone lines in Area i.

Similarly, the fraction with characteristic c in the area served by difficult equipment is estimated by

$$f_c = \frac{\sum_{i=1}^{N} n_{ic} \left[1 - \frac{m_{ie}}{m_i}\right]}{\sum_{i=1}^{N} n_i}.$$

CHAPTER 3

The Questionnaire, Calling, and Billing Data

Belinda B. Brandon

3.1 Introduction

The motivation has already been discussed in Chapter 1 for collecting data on the demographic characteristics of households so that those characteristics can be associated with telephone usage. The present chapter focuses on the design and implementation of the questionnaire for gathering the demographic data. This chapter also describes the calling and billing data that were gathered for the sample customers. After this introductory section comes a discussion of the design of the questionnaire — why particular questions were chosen and how these questions met or did not meet our expectations for desired data. The third section of the chapter contains a description of how the questionnaire was administered to the sampled customers. The fourth deals with the completion ratio — that is, completed interviews to attempted interviews — for the total sample, along with completion ratios for the individual Central Office Areas sampled; it also contains a presentation of response rates for individual questions. In the fifth section is an outline of the procedures that were used to check whether the questionnaire data were accurate and consistent, as well as correctly coded and keypunched. The sixth section describes in detail the calling and billing data that were available for each customer during each of three time periods. It includes a discussion of missing data and of any substitutions of data. The final section is a short conclusion.

3.2 Questionnaire Design

The questionnaire, which is attached as Appendix 3.A, consists of three main sections. The first section contains "Subscriber Attitude Measurement" questions. The second portion of the questionnaire gathers data on the respondent's perception of the calling patterns from his household's phone. The third and most important portion, from our viewpoint, gathers data on demographic characteristics for the individual respondent, such as income, race, occupation, household structure, education, and mobility. For most of these demographic questions, the American Telephone and Telegraph Company's Market Research Information System 1972 questionnaire was used as a model.

The questionnaire begins with five Subscriber Attitude Measurement questions. The first two questions deal with a respondent's perceptions of his telephone service. The third asks whether he feels that he gets his money's worth out of his telephone service. Whether he perceives that its price has gone up less than the rate of inflation is the concern of the fourth. Then the fifth asks whether he feels that he is billed accurately for the number of message units he uses.

There were three reasons for including Subscriber Attitude Measurement questions. Firstly, since many of the subsequent questions were rather personal, it was hoped that these introductory questions would put the respondent at ease. Secondly, it was felt that his answers to these questions might be associated with his answers to the rest of the questionnaire. In fact, a respondent who was white or who had a high income tended to express more favorable opinions of his telephone service than did others.[1] A third reason for the inclusion of these questions was to see how well the answers for this sample agreed with the answers to similar questions in contemporaneous samples of the Subscriber Attitude Measurement project.[2]

---

[1] This observation could be due to factors such as high income persons' subscribing to better service, being better informed, or having higher opinions in general of institutions such as the Telephone Company. It also turns out that answers within the first section correlate strongly with one another.

[2] As an example, a Subscriber Attitude Measurement report found that in March, 1973, 87.4 percent of the customers they sampled in Chicago considered their telephone service to be excellent or good, rather than fair or poor. The result of the same question for the original sample of the present study was that, for the same month as above, 84.0 percent considered their service to be excellent or good. Assuming binomial distributions for the two samples, these observations are not significantly different at the 5 percent significance level.

The second section of the questionnaire contains five questions pertaining to the respondent's use of his telephone service and three questions concerning the type and amount of equipment to which he subscribes. The first question in this second section asks him how many calls he thinks that his household makes each day to people within the city of Chicago. The next question asks him what percentage of his total calls are placed to people within the city of Chicago. Questions eight and nine attempt to elicit his willingness to use his telephone for both local and long distance calls. In question ten the respondent is requested to estimate the average duration of his calls made to people within Chicago. The next question inquires how many telephone instruments he has in his home, and the twelfth how many telephone lines he has. Whether or not he has Touch-Tone® is asked in number thirteen, the last in the second section of the questionnaire.

The answers to the questions that ask the respondent to estimate his own local telephone usage were compared to actual conversation times and numbers of calls made by the same respondents for October, 1972, April, 1973 (near March, 1973, when the questionnaire was administered), and April, 1974. The respondents tended to overestimate both the number of calls and the average duration of calls (see Section 5.4.4 of Chapter 5). It is unclear to us whether the customer does not have a good perception of his number of calls or whether that particular question was misleading. In May, 1974, the Illinois Bell Telephone Company (IBT) began printing on each customer's monthly statement the number of message units actually used by him in a given billing period. This additional information may improve a respondent's estimate of his telephone usage. Misunderstanding the question does not seem to explain why a respondent overestimates his average duration of calls.

The inquiry about any additional lines that were billed separately was made to enable us to include all his telephone usage in analyzing household calling patterns. Since inferences about the number of telephone instruments and Touch-Tone® service can be drawn from the telephone bill, these questions were included to check the accuracy of the respondent's answers. The respondent always answered correctly on whether or not he had Touch-Tone®, but he did not always agree with telephone records on the number of instruments installed (where there was disagreement, the customer usually listed more than did the telephone records). The disagreement could be due to several reasons:

(1) He may not have counted correctly.

(2) He could have counted non-Bell extensions.

(3) Telephone records may not have been correct.

The last section of the questionnaire elicits information from a respondent on demographic characteristics. The content of the questions is listed below, and then the reasons are given for their inclusion.

The first question in this demographic section — number fourteen — determines whether the person being interviewed is the head of the household. Question fifteen asks the respondent whether he owns or rents his present residence, and sixteen asks him in what type of building his residence is contained. Question seventeen inquires how long the respondent has lived at his present address; eighteen, how long he has lived in Chicago; and nineteen, how many times he has moved in the last five years. Question twenty attempts to discern whether the respondent may occupy another dwelling during part of the year. The twenty-first asks him to list, by age group, all household members and to indicate the sex of these members. Next, in question twenty-two, the respondent is requested to indicate how many of these household members are present during each of the four seasons of the year. Questions twenty-three, twenty-four, and twenty-five inquire about the sex, age, and marital status of the head of the household. For both the head of the household and the spouse, education and employment status are asked by questions twenty-six and twenty-seven, while twenty-eight inquires whether each of them does business at home. The next two questions — twenty-nine and thirty — ask for the industry in which the head is employed and for his position. Question thirty-one requests the respondent to indicate which of several income groups best represents his family income. He is asked in the final two questions to indicate his ethnic background, then age and sex (for brevity, in most of the remainder of the book, ethnic background is referred to as "race"; furthermore, no distinction is made between the race of the respondent and the race of the rest of the household).

There were two principal reasons for the inclusion of the above demographic questions: Firstly, all but two of them inquire about variables that seemed as if they might have some effect on telephone usage. Of these two exceptions, one asks whether the respondent is the head of the household, and the other asks for the respondent's age and sex. Secondly, a set of variables were thought to be well related to income, so that, using them, one could estimate income for respondents who refused to answer the question on income. The information on these variables is obtained by the questions on sex, age, marital status, and occupation of the head of the household; education and employment status of the head and spouse; and race.

Certain questions were included for other reasons in addition to or instead of the above. It was assumed that a head of a household was likely to be more knowledgeable about the demographic characteristics of his household than others of its members. The respondent was asked whether he was the head of the household early in the crucial part of the interview, so that in the case of a negative answer, arrangements could be made for contact with the head. In practice, contact with the spouse was accepted. Since a spouse could frequently provide as accurate information as a head, this question could have been phrased better. Also, some female respondents were upset since there was only provision for one head. The revised version of the questionnaire in Chapter 11 substitutes the classifications of male head and female head for head and spouse; and it asks all the other demographic questions for both the female and the male heads of the household.

The question on duration of time at the present address was used as a check on the accuracy of the data. As discussed in the next section, the original sample was chosen in October, 1972, at which time calling and billing data were first collected; but the questionnaires were administered later. If a respondent indicated that he had lived at his present address less than six months, then we verified that the customer who was interviewed was the same as the customer who was originally selected for the sample. The final question, on the age and sex of the person filling out the questionnaire, was included because some studies have indicated that answers to some questions are affected by these characteristics.

3.3 Questionnaire Administration

It was decided to administer the questionnaire by personal interview.[3] This decision was made for two reasons. Firstly, it was felt that asking the respondent his ethnic background in either a mailed or telephoned questionnaire would be too sensitive, but a personal interview enables it to be determined by direct observation.[4] Secondly, a personal

---

[3] Opinionmetrics, a Chicago opinion research firm, was retained by BTL to administer the questionnaire. The instructions to the interviewers were handled by R. J. Hevrdejs, then of IBT's Management Sciences. Also, Management Sciences instructed the agency in the method of coding the questionnaire responses.

[4] The question on race was not in the original version of the questionnaire that was given to each sampled respondent, although the interviewer was instructed to note the race of the respondent.

interviewer might help the respondent who could not adequately read English to complete the interview, while such help would not be possible in the case of a mailed questionnaire.

Each of the customers in the sample was sent a copy of the questionnaire. He was asked to look over the questionnaire and to retain it until an interviewer contacted him. When an interviewer visited the customer's residence, if the customer had not filled out the questionnaire already, then the interviewer would wait for him to do so or, if necessary, help him to complete the questionnaire. After completing the visit, the interviewer recorded the ethnic background of the respondent. If a respondent refused to fill out the questionnaire because the requested information was too personal, then he was encouraged at least to complete the sections on telephone service and usage.

The interviewing agency had agreed that, for each of the sampled customers, it would attempt to effectuate a personal contact three times, if necessary. But after the agency satisfied this agreement, only 365 interviews out of the sample of 717 had been completed. In less than half the instances, this low response rate was due to outright refusals; most problems were caused by respondents' either not being at home or refusing to open their doors to someone they did not know. The agency, in order to protect the safety of its employees, only interviewed until about six o'clock in the evening. Interviewing only during the day made it difficult to contact customers who worked.

The decision was made to attempt to contact by telephone twenty-five of the sampled customers whom the agency had been unable to contact personally. For these twenty-five customers it was decided to ask the ethnic background question. Directly asking this question resulted in no refusals and no irate customers. It should also be noted that the agency charged approximately three-quarters less for a completed telephone interview than for a completed personal interview. Therefore, it was decided to contact by telephone the remaining customers in the original 717 sample whom the agency had as yet been unable to contact. Three attempts to contact each customer were made, if necessary. By this method, 148 additional interviews were completed. In total, then, 513 out of 717 interviews were completed.

In April, 1973, three #1 ESS machines were introduced into the Beverly and Pullman Central Offices. At that time it was decided to administer the same questionnaire to a sample of customers from these three ESS prefixes. For this sample, 132 customers were drawn. These 132 were sent the version of the questionnaire that is attached as an appendix. As an experiment, they were asked to fill out the questionnaire and return it by mail to IBT. Each questionnaire was coded with the respondent's telephone number so that the demographic data could be related to billing and calling data gathered for the same respondent. Sixty-six questionnaires were returned by mail.[5] It was decided to attempt to contact the remaining 66 nonrespondents by telephone. Three attempts to contact each of the nonrespondents were made, if necessary. An additional 44 interviews were completed in this way, for a total number of completions of 110 out of 132.

Note that the questionnaire was administered in mid-1973, whereas calling data were not only collected in April, 1973, but also in October, 1972, and April, 1974. An attempt to correlate questionnaire responses with calling data from the latter two months may then involve a few errors. For example, by April, 1974, there were no households in the sample who had lived in Chicago less than one year, even though that response is recorded for some households in 1973. Further, the questionnaire requests information on customers' family incomes for 1972, which measures with some error their incomes in 1973 and especially 1974. A few other problems, which are relevant only to particular analyses, are mentioned in the appropriate chapters.

3.4 Response Rate

Let us define a respondent as having "completed" a questionnaire if he answers questions beyond the first section, on telephone service. According to this definition, questionnaires were completed by 513 of the original 717 sampled customers, for a 72 percent response rate. Of the 204 customers for whom we failed to obtain completions, 89 were contacted but for various reasons declined to cooperate. There still remained 106 whom we were unsuccessful in contacting after three attempts to visit the respondent personally, and three attempts to reach him by telephone. We were unable to interview seven respondents due to a language barrier (none of these were Spanish speaking). An additional two customers disconnected their telephones between October, when the sample was originally drawn, and March, when the questionnaire was administered.

---

[5] A questionnaire was sent to each of these customers with a stamped envelope addressed to IBT. Several of the completed questionnaires were mailed by respondents after they had removed the stamps, so that postage due had to be paid. Interestingly, the respondents who removed the stamps tended to report higher incomes than those who did not; their reported higher incomes tended to be consistent with their occupations.

One hundred ten out of the 132 customers who were sampled in the Beverly and Pullman ESS exchanges completed interviews, for a response rate of 83 percent. Of the 22 customers who failed to complete interviews, 10 refused and 9 were unavailable. Two customers disconnected their service between April, when the sample was drawn, and May, when the questionnaire was administered. One respondent was unable to complete an interview due to a language problem.

The overall response rate for the original Chicago sample plus Beverly and Pullman was 73 percent — that is, 623 completed interviews out of 849 attempted interviews. Exhibit 3.1 shows, by Central Office Area, the number of questionnaires sent out, the number of actual completions, and the completions as a percentage of the number of questionnaires sent out. According to this exhibit, Wabash had a response rate of 100 percent, but there was only one customer in its sample. Kedzie was the Central Office Area with the next highest response rate — 90 percent. The response rates of the other Central Office Areas ranged down to 62 percent for Hyde Park.

It turns out that the customers with "completed" questionnaires have different local and suburban telephone usage rates from the sampled customers without "completed" questionnaires. Exhibit 3.2 enumerates the sample means of various measures of April, 1973, usage rates for these two groups, along with the z-statistics for a Normal test for the difference between each pair of means. The differences are statistically significant at the 5 percent level or better in the cases of the number of local calls, total local conversation time, total message units, and toll charges (by a two-tailed test).[6] These differences do not necessarily imply that there is a bias in any estimated relationship between a measure of telephone usage and demographic characteristics. If the difference in usage between those who did and did not complete the questionnaire is simply due to the two groups' having, say, a different income mix, then the estimated effect of income is unbiased. There is a problem, however, if, given the demographic characteristics entering the model, the probability of questionnaire completion is

---

[6] For all these comparisons except for toll charges, the few customers are excluded who receive concessions from parts of their bills by virtue of their being telephone company employees. (See footnote 10 below.) Three particular customers are also excluded from the above toll comparisons who are excluded from the analyses of Chapter 10. If these customers are included, then the results for toll charges are as follows: with completed questionnaires $6.21, without completed questionnaires $8.44. The z-statistic is 1.70, which is not significant at the 5 percent level.

correlated with telephone usage. Investigating to what extent there is such a correlation would be very difficult and has not been done. This is a problem associated with almost every survey.

Exhibit 3.1

Completed Questionnaire Percentages

Central Office	Number Sent Out	Number of Completions	Percentage of Completions
Interview In Person and By Telephone			
Wabash	1	1	100
Kedzie	11	10	91
Monroe	119	103	87
Prospect	31	26	84
Portsmouth	114	85	75
Dearborn	16	11	69
Merrimac	42	29	69
Lawndale	3	2	67
Calumet	203	134	66
Superior	156	99	63
Hyde Park	21	13	62
Total	717	513	72
Interview By Mail and By Telephone			
Beverly	62	52	84
Pullman	70	58	83
Total	132	110	83
Overall Total	849	623	73

Exhibit 3.2

Monthly Usage Rates of Sampled Customers With and Without Completed Questionnaires

Measure	With Completed Questionnaires	Without Completed Questionnaires	z-Statistic
Number of local calls	103.9	83.2	3.04
Average duration of local calls (minutes)	6.2	5.8	0.88
Total local conversation time (minutes)	605.7	465.2	2.91
Suburban message units	63.9	53.0	0.99
Total local and suburban message units	167.7	136.2	2.29
Total charges ($)	21.10	22.85	1.06
Vertical service charges ($)	1.58	1.60	0.06
Toll charges ($)	5.67	8.44	2.22

The refusal rates for each individual question are in Exhibit 3.3, for the combination of the original sample and Beverly and Pullman. Only three questions had a refusal rate greater than four percent (or twenty-five out of 623): those requesting information on the spouse's employment status, on whether the spouse does business at home, and on family income; their refusal rates were 5.0 percent, 4.0 percent, and 16.1 percent, respectively.

3.5 Error Checking

After the interviews were completed and the questionnaires were turned over to IBT's Management Sciences, the latter checked a sample of the questionnaires to ensure that the agency had followed the coding instructions properly. Then the questionnaires were sent to Bell Telephone Laboratories (BTL) for further error checking and analysis.[7]

---

[7] In this section the original sample and Beverly and Pullman are not distinguished since there was no difference in the error checking procedure.

Exhibit 3.3

Refusal Rates on Individual Questions

Question	Number of Refusals (Out of 623)	Percentage of Refusals (Out of 623)
The quality of service	0	0.0
Service problem concern	1	0.2
Pleasant to deal with	5	0.8
Getting money's worth	2	0.3
Increase in Telco cost	5	0.8
Message unit bills okay	17	2.7
Average daily local calls	14	2.2
% of calls in Chicago	9	1.4
Using phone, local calls	2	0.3
Using phone, long distance	0	0.0
Average local call length	9	1.4
Number of phones in home	1	0.2
Number of lines in home	0	0.0
Touch-Tone® phone in home	0	0.0
Are you head of house	5	0.8
Own or rent residence	3	0.5
Type of building	2	0.3
How long at this address	4	0.6
How long in Chicago	4	0.6
Moves in past five years	5	0.8
Own or rent two houses	4	0.6
Members of family	7	1.1
People here during year	11	1.8
Sex of head of household	4	0.6
Age of head of household	22	3.5
Marital status of head	11	1.8
Education of head	13	2.1
Education of spouse*	18	3.0
Employment of head	12	1.9
Employment of spouse*	31	5.0
Business at home, head+	3	0.5
Business at home, spouse*,+	25	4.0
What type of business+	23	3.7
Title of head+	23	3.7
Family income for 1972	100	16.1
Race of respondent	9	1.4
Age of respondent	20	3.2
Sex of respondent	4	0.6

* If married.
+ If employed.

We checked all 623 completed questionnaires for correct coding and keypunching. In addition, we made several consistency checks as follows: For any respondent who claimed to have more than one telephone line, we asked IBT's Commercial Department for verification. If the respondent in fact did have more than one line, then we combined the usage data from the additional line or lines with the usage data from the sampled line. If, however, the respondent had only one line, then we corrected his response. If a respondent claimed to subscribe to Touch-Tone®, then we checked this claim with the bill; there seemed to be no unexplainable inconsistencies. If a respondent had lived at his address less than six months, then we made sure that the respondent who had filled out the questionnaire was the same as the one for whom we had collected billing and calling data in October, 1972. We checked for consistency between answers on questions 15 and 16; for example, if a respondent answered that he rented an apartment, then we made sure that his type of residence was not coded as a one-family house. We checked to ensure that the respondent lived at least as long in Chicago as he had lived at his present address. Further, if he had lived at his present address for more than five years, then we verified that he had not said that he had moved in the past five years; or if he had lived at his present address for less than six years, we made sure that he had moved at least once in the past five years. For the questions on number of moves and length at the present address, we attempted to resolve any inconsistencies by reference to IBT's Commercial records; then we changed the answers to be consistent with the latter records. If there were no person of the head's sex and age (or close to it) entered in the table for household composition, then we added that information in the table (when interviewed, heads sometimes neglected to include themselves). In those cases where we added the head to the table for household composition, one person was also added to the number of persons reported for each season if the maximum number for the four seasons fell short of the corrected sum for household composition. There were also two cases in which there was no problem with the head, but the numbers recorded for the seasons fell short of the sum for household composition; since in these cases there were a large number of children, it was assumed that the respondent miscounted, so the number for each season was increased by one. In cases in which the head of the household was listed as married and female, we rejected that answer if she were unemployed and the male spouse was employed; *i.e.*, we changed the head of the household to male in order to be consistent with the treatment in the rest of the questionnaires. If the marital status of the head of the household were listed as other than married, then we ensured that there was no answer to spouse's education, employment status, nor business at home. If the head of the household were listed as employed full- or part-time, then we confirmed that the business and position of the head of the household were both listed as other than unemployed. If the head of the household were coded as being unemployed or retired, then both the business and position were also supposed to have "unemployed" as their entries. If either the head of the household or spouse were listed as unemployed or retired, then we checked to ensure that there was no answer for business at home.

3.6 The Calling and Billing Data

An important part of the data was about the telephone calls of the sample of Chicago customers. For the purpose of this book, a "call" denotes a completed call, excluding any attempts that are incomplete by virtue of their resulting in a busy signal, no answer, abandoned dial, *etc.* The data also exclude calls to directory assistance. For the subsequent discussion, recall the terminology introduced in Chapter 1 that a "local" call both originates and terminates within Chicago; a "suburban" call originates in Chicago but terminates in a specified band of Illinois suburbs; and a "toll" call is one between Chicago and the area outside that band of suburbs.

Below is the description of the kind of data collected for October, 1972, April, 1973, and April, 1974. Exhibit 3.4 summarizes the sample size for each type of data for each date, broken down into the number of customers who did or did not respond to the questionnaire.

For the billing periods[8] that began in October, 1972 (or in some cases adjacent months), data were collected for each customer on the total local conversation time and total message units, total number of local calls, and the numbers of local calls per duration period (1 minute, 2 minutes, ..., 20 minutes, over 20 minutes). Summary bills for individual customers in the sample were also collected. The information contained in those bills are as follows: whether the customer was a residence or business, message unit allowance, total bill, balance forward from the previous month, "basic plus vertical service charges", message units in excess of the allowance, total toll charges, and other

---

[8] In order to even out its accounting load, IBT divides its customers into ten "billing periods". One group of customers is billed for the interval of the first day of a month through the last, another for the 4th of the month through the 3rd of the next month, another for the 7th of the month through the 6th of the next month, *etc.*

charges and credits.[9] For the 717 customers in our original sample, calling data were available for 696 and billing data for 691. Because of data loss, calling data from September were substituted for October in the case of 80 customers; November data were substituted for 119 customers. No substitutions were necessary for billing data. Of those for whom data were available, 6 benefited from "concessions" because they worked for the telephone company; these customers are excluded from most analyses in this book, since their concessions effectively provided discounts for certain services.[10] However, the customers with concessions are included in Chapter 10.

Exhibit 3.4

Amount of Data of Each Type for Each Date

Data	Question-naires	No Question-naires	Total
October, 1972			
Summary Calling	497	199	696
Summary Billing	494	197	691
April, 1973			
Summary Calling	585	207	792
Detailed Local Calling	331	136	467
Billing	593	206	799
April, 1974			
Detailed Calling	424	155	579
Summary or Detailed Calling	460	164	624
Billing	483	175	658
Detailed Toll	465	166	631

---

[9] "Basic service charge" is the monthly fee for a line, a single plain telephone instrument, and the customer's choice of number of message units as an allowance. "Vertical service charge" is the monthly fee for extras such as Touch-Tone®, extensions, fancy telephones, *etc.*

[10] The telephone employees with these so-called "Class B" concessions receive a 50 percent discount on their local service charges (basic service charge and charges for message units in excess of the allowance) and on certain vertical services, but not on toll. A "Class A" concession, screened out of the sample at the beginning, confers more benefits to the employee.

The collection of the October data, originally viewed as a pilot run in which experience could be acquired, was very successful. But it was decided that more detail was desirable. Further, two additional areas began to be served by #1 ESS machines, so a sample of 132 residence customers was drawn from the new areas. IBT attempted to gather billing data and certain detailed local calling data for all sampled customers for billing periods beginning in April, 1973. The possible sample size was then a total of 849. Of these, summary calling data were available for 792 and billing data for 799. Also, detailed data on local calls — times of connection and disconnection, date, and called prefix — were collected on 467 of the sample customers. The April, 1973, detailed calling data are not used in the reported analyses because the sample is too small and/or is limited to a confined area of Chicago; however, billing data and summary calling data are used in some chapters. July bills were substituted for April in the case of 17 customers, and May detailed calling data were substituted for April for 81 customers. In the 1973 data set, 9 of the customers received telephone company employee concessions.

IBT collected additional detailed calling data on the sample customers for billing periods beginning in April, 1974, including data on all outgoing local and suburban calls, except to directory assistance. Summary calling data were also collected, but they are only employed in the regression analysis in cases for which detailed data are unavailable; the detailed data are used for the analyses in Chapters 5, 7, and 8. Billing data were collected for billing periods beginning in February, March, and April of 1974. For billing periods beginning in April, detailed data on individual toll calls were also collected — their times of connection and disconnection, date, area code and prefix of the other party, and type of call (*i.e.*, direct dial, collect, credit card, third party, or person-to-person). August summary and detailed calling data were substituted for 26 customers; June billing data were substituted for April billing data for 65 customers. Out of a possible 849 customers, detailed local and suburban calling data were accumulated for 579, complete summary calling data for an additional 45 (so the number with detailed or summary data totaled 624), and some partial calling data for an additional 20 (for a grand total of 644). At least April billing data were collected for 658, and the number of customers with valid detailed toll data or with no toll calls totaled 631. Eight of the sample customers received telephone company employee concessions.

Since so much of the book's analysis uses the 1974 data, it is useful to see whether there is a potential problem due to missing data. Of the customers for whom some 1974 data were missing, we know 111 of

44 The Effect of Demographics on Telephone Usage

them to have been disconnected either at their own request or for non-payment; almost all of an additional 76, for whom we have neither calling nor billing data for 1974, were also probably disconnected. Excluding from consideration the customers with concessions, there are also 81 customers whom we knew to be connected but for whom detailed local and suburban calling data were missing. For the set of customers for whom 1973 summary calling data were available, we compared the 1973 telephone usage of the customers for whom we have 1974 detailed local and suburban calling data to that of the customers for whom we do not.

Exhibit 3.5

Mean Monthly 1973 Telephone Usage of Sampled Customers With and Without 1974 Data

Measure	With 1974 Data	Without 1974 Data	z-Statistic
Number of Local Calls	95.9	104.8	1.25
Average Duration of Local Calls (minutes)	6.3	5.6	1.90
Total Local Conversation Time (minutes)	573	558	0.33
Suburban Message Units	60.7	61.8	0.07
Total Message Units	157	167	0.60
Toll Charges ($)	5.65	9.07	2.72

The results are shown in Exhibit 3.5.[11] At the 5 percent significance level, one cannot reject the hypothesis that the 1973 local or suburban usage of the groups with and without 1974 data are the same. (The average durations of the two groups are significantly different at the 10 percent level, however.) Of those for whom total toll charges were available on 1973 bills, Exhibit 3.5 indicates that these charges were significantly higher for the customers without valid 1974 data on toll detail than for the others. (Customers who made no toll calls are classed as having toll detail.) It may also be useful to perform

---

[11] No comparisons are shown for billing measures other than toll, since the analyses of bills in Chapter 9 use 1973 data. Customers with concessions are included in the toll comparison.

a comparison for the subset of customers who made at least one toll call in 1973, since much of Chapter 10 analyzes that subset: the 1973 mean toll charges for the set with 1974 data is $10.61, and for the set without 1974 data is $14.77. The corresponding z-statistic is 2.25, indicating a significant difference at the 5 percent level. We have not investigated whether these observed differences are due to the groups' having different demographic characteristics or due to their having different usage for given demographic characteristics.

3.7 Conclusion

On the whole, the questionnaire was successful in gathering the desired data. Suggestions for improvement, however, are made in Chapter 11.

As this exploratory study proceeded, decisions were made to collect calling and billing data that were successively more detailed and complete. Since it has been found that the kind of data taken last was the most useful, I recommend that any similar studies in the future collect data with a very high degree of detail and completeness.

Appendix 3.A

Mailed Version of the Questionnaire

THE
TELEPHONE COMPANY
WOULD LIKE YOUR ANSWERS

We are interested in your opinions as a telephone user so we may improve your service. Perhaps you will help us by answering the questions in this booklet.

You can answer most of the questions by putting a check mark in the square ☐ opposite the answer which comes closest to what you think.

Will you please answer **all** of the questions. A postage paid and self-addressed envelope is provided for your reply.

ABOUT YOUR SERVICE! 8-1

1. How would you rate the quality of telephone service you are now getting?

Poor	☐ 1	9
Fair	☐ 2	
Good	☐ 3	
Excellent	☐ 4	

2. How would you rate the Telephone Company on:

	Poor	*Fair*	*Good*	*Excellent*	
Being concerned about the telephone service problems of the individual customer?	☐ 1	☐ 2	☐ 3	☐ 4	10
Making it easy and pleasant to do business with the company?	☐ 1	☐ 2	☐ 3	☐ 4	11

3. Do you feel that you get your money's worth out of your telephone service?

Definitely Not	☐ 1	12
Probably Not	☐ 2	
Probably Do	☐ 3	
Definitely Do	☐ 4	

4. Do you feel that the cost for telephone service has gone up more, less or about the same as the cost of other goods and services in the past few years?

Much More	☐ 1	13
More	☐ 2	
About the Same	☐ 3	
Less	☐ 4	
Much Less	☐ 5	

5. Do you feel you are generally billed accurately for the number of message units used?

Yes, billed accurately	☐ 1	14
No, billed for **more** than used	☐ 2	
No, billed for **less** than used	☐ 3	

ABOUT YOUR TELEPHONE AND HOW YOU USE IT

6. On an average day, about how many calls do you and others living here make to people **within** the city of Chicago?   _____ calls   15
   (your best estimate)

48   *The Effect of Demographics on Telephone Usage*

7. Of **all** the telephone calls you and others living here make (including long distance), about what percent of these calls are to people living **within** the city of Chicago? Would you say:

Percent of **all** *calls which are to others in the city of Chicago*

25% or less	☐ 1	16
26% to 50%	☐ 2	
51% to 75%	☐ 3	
76% to 89%	☐ 4	
90% to 100%	☐ 5	

8. How do you feel about using your phone for **local calls**? Would you say you use it:

Whenever you feel like it	☐ 1	17
Whenever you feel like it — within reason	☐ 2	
Only when it seems fairly important	☐ 3	
Only when it's absolutely necessary	☐ 4	

9. How do you feel about using your phone for **long distance calls**? Would you say you use it:

Whenever you feel like it	☐ 1	18
Whenever you feel like it — within reason	☐ 2	
Only when it seems fairly important	☐ 3	
Only when it's absolutely necessary	☐ 4	

10. On calls made by you and others living here to people in the city of Chicago, would you say that the average length of these calls is:

Under 5 minutes	☐ 1	19
5 to 10 minutes	☐ 2	
Over 10 minutes	☐ 3	

11. How many phones do you have in your home? _____ 20
    (Please specify number)

12. How many telephone lines do you have in your home? Each line has a different telephone number for example _____ is one line and 555-3398 would be another. (Most customers have one line.) _____ 21
    (Please specify number)

13. Do you have TOUCH-TONE® (Pushbuttons) in place of a dial on your telephone?   Yes ☐ 1   22
                                                                                    No ☐ 2

ABOUT YOUR FAMILY AND HOME

14. Are you the head of the household?   Yes ☐ 1   23
                                          No ☐ 2

15. Do you or other members of your household own or rent your present residence?

    Own: *Rent:*
    House   ☐ 1   24   Apartment   ☐ 1   25
    Cooperative   ☐ 2   House   ☐ 2
    Condominium   ☐ 3   Room in a house   ☐ 3
    Multi-Family ☐ 4   Room in an apartment ☐ 4
                              Other   ☐ 5

16. In what type of building is your residence located?

One Family House	☐ 1	26
Two Family House	☐ 2	
Three or more Family (Apartment Building)	☐ 3	
Mobile or trailer home	☐ 4	

17. How long have you lived at this address?

Under 6 months	☐ 1	27
Six months to 1 year	☐ 2	
1-2 years	☐ 3	
3-5 years	☐ 4	
6-10 years	☐ 5	
11-15 years	☐ 6	
More than 15 years	☐ 7	

18. How long have you lived in the city of Chicago?

Under 1 Year	☐ 1	28
1-2 years	☐ 2	
3-5 years	☐ 3	
6-10 years	☐ 4	
11-15 years	☐ 5	
More than 15 years	☐ 6	

19. How many times have you moved in the past five years?

None	☐ 1	29
Once	☐ 2	
Twice	☐ 3	
Three times	☐ 4	
Four or more times	☐ 5	

20. Do you (or any members of your household) own or rent a second home or other living quarters which you occupy during part of the year?

Yes	☐ 1	30
No	☐ 2	

21. Please fill in the table below for **all** family members and others who normally live in your home at least 2 months a year, including yourself. For each age group, write the number of males and females normally living at home.

Age Group	Males		Females	
Under 10	_____	31	_____	40
10 to 12	_____	32	_____	41
13 to 15	_____	33	_____	42
16 to 18	_____	34	_____	43
19 to 24	_____	35	_____	44
25 to 34	_____	36	_____	45
35 to 54	_____	37	_____	46
55 to 64	_____	38	_____	47
65 and over	_____	39	_____	48

22. How many people live here during **each** of the seasons of the year?

Please record the number living here for each season

	NUMBER	
Summer	_____	49
Fall	_____	50
Winter	_____	51
Spring	_____	52

23. Is the head of the household male or female?

Male	☐	1	53
Female	☐	2	

24. What is the age of the head of the household?

_____ years   54
                   55

25. What is the marital status of the head of household?

Single	☐	1	56
Married	☐	2	
Separated	☐	3	
Divorced	☐	4	
Widowed	☐	5	

PLEASE ANSWER THE FOLLOWING QUESTIONS ABOUT THE HEAD OF THE HOUSEHOLD *AND* ALSO ABOUT THE HUSBAND OR WIFE (SPOUSE) OF THE HOUSEHOLD, IF APPLICABLE.

26. What is the highest grade attended or degree received?

	(Check one)	(Check one, if applicable)
	HEAD	SPOUSE
Eighth Grade or Less	_____ 57	_____ 64
Some High School	_____ 58	_____ 65
High School Graduate	_____ 59	_____ 66
Some College	_____ 60	_____ 67
College Graduate	_____ 61	_____ 68
Some Graduate School	_____ 62	_____ 69
Graduate Degree	_____ 63	_____ 70

27. What is the employment status of the head and spouse of the household?

	(Check one)	(Check one, if applicable)
	HEAD	SPOUSE
Employed full time	_____ 71	_____ 76
Employed part time	_____ 72	_____ 77
Unemployed looking for work	_____ 73	_____ 78
Unemployed not looking for work	_____ 74	_____ 79
Retired	_____ 75	_____ 80

28. If "employed" do the head or spouse of the household use the home as a place of business?   *Use home as place of business?*   8-2

9

(Check if applicable)

	HEAD		SPOUSE		
Yes, all the time	☐	1	10 ☐	1	11
Yes, occasionally	☐	2	☐	2	
No	☐	3	☐	3	

29. In what type of business, industry or service is the head of the household's company or employer engaged? Please be specific, *e.g.*, steel manufacturing, education, textile wholesale, state government, hardware retailing, *etc.* If the head of household is self-employed, please state.

_____

12
13
_____
14
15

30. What is the title or position of the head of the household? (*e.g.*, foreman, doctor, carpenter, teacher, saleslady, machinist, *etc.*)

16
17
_____
18
19
20

31. Would you please check off your combined approximate family income for *1972* (that is, total wages, interest, dividends, *etc.*, of all family members presently in your household before taxes and other deductions).

Under $5,000	☐	1	21
$5,000 — $8,999	☐	2	
$9,000 — $14,999	☐	3	
$15,000 — $19,999	☐	4	
$20,000 — $29,999	☐	5	
$30,000 or more	☐	6	

32.	Ethnic background of person filling out questionnaire.	Black	☐ 1	22
		Oriental	☐ 2	
		Spanish	☐ 3	
		White	☐ 4	
		Other	☐ 5	
33.	Age and sex of person filling out questionnaire.	_____ Age		23
		Male	☐ 1	24
		Female	☐ 2	

THANK YOU!

CHAPTER 4

The Data Base Design and Implementation

Robert McGill

4.1 Introduction

This chapter contains a description of the data base used in the Chicago residence usage study, the philosophies underlying its development, and the mechanisms and programs used to access the data.

There are two motivations in writing this document. First, it is intended to document the structure and means of accessing the data base. Second, it is hoped that it may aid others facing a similar project by providing a description of the types of planning and strategies that underlie the generation of any successful data base. In this latter regard, it should be pointed out that there is no single "correct" approach to managing data, and that what is set forth here is, in large measure, a reflection of the opinions and prejudices of the author (which are many and strong).

4.2 Overview of the Data

The questionnaire subdivided the sample into two groups; those who replied to some (or all) of the questions beyond the section on attitudes toward telephone service, and those who did not. Of the 717 customers originally chosen, 513 responded to at least one question in addition to those on attitudes; 204 provided no such response. When reduced to machine sensible form, the questionnaires resulted in 69 items of data (including the telephone number) for respondents, and 2 for non-respondents (telephone number and reason for no response).[1] A substantial amount of data was missing due to respondents' refusing to answer one or more questions. Hence after only the first step, one was faced with the twin problems of non-uniformity in structure and missing data. The number of individual data items plus indicator flags was approximately 35,000.

The second data type, obtained for each customer in the sample, was the total number of local calls, total local conversation time, total message units, and the number of one-minute local calls, two-minute calls, ..., 20-minute calls, and the number of calls longer than 20 minutes. These data were provided initially for October, 1972, and April, 1973, and, except for total message units, applied only to local (intra-Chicago) traffic. The number of individual data items then exceeded 70,000.

The third data type consisted of each customer's bill (excluding toll detail) for October, 1972, and April, 1973. Each bill provided 12 data items (although many instances of missing data were encountered). Data items then exceeded 85,000 in number.

The last set of data available at that time consisted of complete detail (date, time, duration, *etc.*) for each local call made by each sample customer during April, 1973. The number of data items in this set, while not initially known, was estimated at well over a half million (an estimate which proved low).

It seemed likely that even more data would be gathered. Between 1.5 and 2 million data items seemed a reasonable guess. Note that for simplicity the discussion at this stage is limited to the data described above. Indeed this is appropriate, since these are the data for which the data base was designed. Additional data obtained in 1974 are described later.

4.3 Order From Chaos — Imposing Structure

With the exception of the call detail, all original data were initially handled in the form of punched cards. A substantial amount of analysis was performed with the data in that form. Subsequently, much of the data was converted to card-image tapes and files. Yet the various sets of data remained fundamentally disjoint rather than providing the analyst with a cohesive whole with which to work.

---

[1] Note that, for reasons of privacy, neither the name nor the address of the customer is included in the data base.

When the author first became associated with the project, his first priority was to combine the data into a coherent data base. However, a data base implies structure, and none was clearly defined. Further, and this point is too often overlooked, the appropriate structure is generally best defined by the anticipated *use* of the data rather than by the data themselves.

At first glance the data would appear to be somewhat hierarchical on the telephone number:

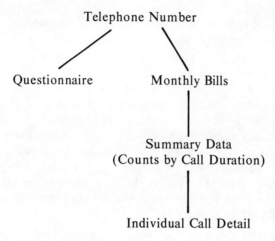

In fact, if the data were intended for use by a business office, this would be the appropriate form. The customer, identified by his telephone number, will generally have a question (or complaint) about a specific month's bill. Records of complaints, requests for payment, *etc.* (similar in concept to our questionnaire), tell the service representative what to expect. Summarized data may suffice to answer the question; if not, the detail of a particular call may be required. While random access of the data (described in Appendix 4.A) might offer certain technical efficiency, sequential organization would be acceptable (and, in fact, is often used in Business Offices in the form of tub files).

In our project, the anticipated use of the data was completely different. Since the aim was to characterize usage in terms of demographic variables, the telephone number, rather than being a necessary point of departure, becomes simply one more piece of data. In fact, other than providing a means of relating the various items of data for a single customer (which may be better done in other ways), it is actually one of the least useful items of data. Let us, however, assign it equal status to the other questionnaire data for the moment.

The pattern which then begins to emerge is one which divides the data into two segments — a fixed-length set of data for each customer (questionnaire responses, counts of calls, bills), and a variable length set of call detail, the size of which is a function of the number of local calls. (The problem of missing questionnaires is being postponed for the moment, not ignored.)

Let us first examine the fixed-length set. Conceptually, we may view this set as a matrix with the customer defining the rows and the responses the columns. Were this matrix small enough (say 5000 items or less), it could probably best be handled as a unit which could be read completely into a program and manipulated at will. Unfortunately, our matrix is far too large to allow this approach (85,000 items and growing). We must subdivide the data before storing it. In the context of the business office example previously given, one would clearly divide by row (customer). Not so in our case. Our use of the data would more frequently call for the use of a single measure for all customers who may be categorized or subselected on another measure. For example, we would very likely wish to look at the distribution of bills categorized by income. We must divide by columns — by single measures for all customers. Further, if the same order is maintained in each set, there is no need to maintain customer identification for each item — it is implicit in the position.

However, to this point we have not defined any customer order. In fact, one may make a case for a large number of orderings, but none is clearly to be preferred. What is critical is to divide the questionnaire non-respondents from those who did respond, thus avoiding the storing (and repeated detecting) of about 14,000 missing entries. Within the two groups, we will put the data in order of telephone number (not necessarily an optimal choice, but one that is intuitively obvious and has little against it).

At this point, we have defined the structure to be imposed on the fixed-length sets: questionnaire respondents separated from the non-respondents, and each measure (or response) in a separate data vector. Contents of the vectors are in telephone number order (and telephone numbers are in one of the vectors). This defines 221 data sets. Later, we shall discuss how each is identified.

Next we consider the call detail — the variable-length data sets. Eleven measures were generated for each call — date, time, duration, *etc.* The number of calls for an individual customer in the data varied from 0 to 572 (*i.e.*, from 0 to 6,292 data items). Treating all the data as an array is obviously absurd. Further, it appeared from a study of the data that one would seldom want all the measures simultaneously. One is led to the concept of storing each measure for each customer in a separate vector. However, since this would generate in excess of 6,000 vectors, it would be best that this fact be transparent to the user; we will discuss this issue shortly.

At last, our structure is defined, but is based on the anticipated *use* of the data rather than on any (apparent) inherent structure. With the number of data sets and the anticipated use, it goes without saying that our access to the data must be random.

4.4 Users and Anticipated Use

The next issue to be resolved in the design of the data base focuses on projected use — how much and by whom? First, the users must be viewed *objectively*. This is *not* the time to be a "good guy" and say everyone is a computer expert; this simply is not true. In the case of the data under discussion, it was determined that all likely users could write Fortran, but that the level of competence in the language ranged from virtual beginner to system programmer level. It was also apparent from the exploratory nature of some of the analyses anticipated that many of the data sets would be retrieved a very large number (at least several hundred) of times. Hence efficiency could not be ignored in this case.[2] The situation encountered is not an uncommon one — the simultaneous need for ease of use and efficiency. Since there are trade-offs, compromises must be made.

Also important, at this stage of planning, is consideration of the likelihood of expanding and/or modifying the data base. Too often great efficiency is gained at the price of almost total inflexibility. In our case, the data base was almost certain to grow but would probably have few changes made to data items or sets once incorporated within the base. This fact would have to be considered.

Human engineering also plays a part. What mechanism will the user prefer in identifying which of the thousands of data sets he or she wants? What should happen if the user makes a mistake?

Basically a data set may be referred to in two ways, by a literal (and hopefully mnemonically reasonable) *name,* or by a *number.* For example, one might refer to the set containing the telephone numbers

---

[2] Designers of data bases frequently are unaware of (or choose to ignore) the fact that a data set which will be read only a few times need *not* be read with great efficiency, and large amounts of time invested in gaining efficiency in such cases are truly wasted.

of our respondent group as "TELNOQ" or as 1. Surprisingly fierce arguments have been known to rage over this seemingly small point. In truth, there is a place for both methods, and it is often moderately easy to provide the user with a choice. This, in fact, was done.

But what should be done if the user makes an error?[3] The author believes, as a matter of dogma, that routines should be designed so that the *only* routine supplied to a user which may ever cause a program abort is a routine designed *explicitly* for the purpose of aborting a program. Clearly, indication of an error condition must be given to the user (perhaps by means of an alternate return from a routine), but the user must *always* be given the final say as to whether he or she wishes to abort or attempt some form of error recovery.

Based on the preceding considerations, it was decided that users of this data base would be provided a set of Fortran-callable subroutines which would allow retrieval of any data set by specifying either a mnemonically reasonable name or a data set number. In case of error, the routines would take an alternate return. Implicit in this plan is the fact that the true structure of the data base (as yet undefined in specific detail) would be totally transparent to the user. Except in the case of a user community comprised entirely of extremely knowledgeable programmers (who presumably will jointly design and implement the base), this transparency is desirable. First, since the user is unaware of the base structure, he or she need make no effort to understand it. Second, those responsible for maintenance of the base may freely modify the structure, constrained only by the consideration that the user must not be aware of change. Finally, and perhaps more importantly, the user will be less inclined (or at least less able) to alter the data base.

4.5 Media, Methods, and Efficiency

This section is intended, in so far as possible, to view the implementation of the data base in a machine-independent fashion. Our requirements are for a rather large amount of storage space (room for over a million entries), random access capability, reasonably good efficiency in data retrieval, relative ease and efficiency in enlarging the data base, and, of course, Fortran compatibility (which will simply be assumed available).

---

[3] The "user" here is considered as being one who may only *read* the data base. Somewhat different rules must apply to those (hopefully few) who may *write* on it.

The requirement for random access restricts the choice of storage medium to either drum or disc files — either permanent disc or on dismountable disc pack. Except in those rare installations where cost is no object, the large amount of storage required will probably indicate that a dismountable disc pack is, from an economic viewpoint at least, our best choice. Additionally, it will provide more than enough space (likely with some left to handle growth), and it is an efficient medium to read.

The choice of random access software is a highly machine- and location-dependent one. At minimum, any machine will provide a bare-bones capability in its assembly language. Hopefully higher level packages will be available. We must beware, however, of choosing one which will provide more capacity than we require. Our requirement is simply to read and write simple data vectors identified by either an alphabetic or numeric identifier. We specifically do *not* have any requirement for elaborate pointer mechanisms such as are found in Honeywell's IDS and IBM's IMS systems.[4] While it might initially seem that the additional capability will do no harm, one must not forget cost. Specifically, both the systems noted above are expensive to initialize or expand, particularly when the number of data sets involved is large. We are not only getting more than we need, we are also paying for it. Yet once again we are faced with a trade-off. Given the choice of IDS or writing a random access system using the machine language I/O primitive functions, would the time required to write a usable system be justified by the dollar saving over IDS? (Programming is also not free — particularly machine language I/O programming.)

Whatever system is finally chosen, it is most important that the inner working of the system be understood if efficiency is to be achieved. Certain types of algorithms (notably those using hash coding techniques, as discussed later) are extremely sensitive in terms of efficiency to the structure of the data set identifier. A small amount of effort here can pay large dividends later.

4.6 Bringing up the Data Base

When all the design aspects are considered and the data base designed, it would appear that only implementation remains. Sadly, this is not true; the most difficult part lies ahead. It is almost axiomatic

---

[4] See Honeywell Information Systems, Inc., "Integrated Data Store", Publication BR69; and International Business Machines Corp., "Information Management System/360, Version 2", Publication SH 20-0910/11/12/14/15.

that any large set of data will contain errors of both commission and omission. We must identify and, where possible, correct them. This, in fact, is often the largest single job in establishing the base. It is also the most critical. Further, the job is often made more difficult by the fact that the ultimate user, often completely naive in the realm of computers, will attempt to collect the data himself to "save you the trouble".

The next step is programming. The author firmly believes that the routines used to *read* the data from the data base should be completed *before* the routines that will write data onto it are begun. This approach offers several advantages. First, minor points in the structure of the data base will not cause the read routines to be poorly designed from a human engineering standpoint. Modifications in the structure of the base may be made instead. Secondly, producing the read routines first insures that the base has been completely designed (for if it has not, you will not know in detail how to read it). Finally, possible simplifications in the original base design may become apparent while writing the read routines.

Clearly the routines used to access the data should be efficiently written since, by the very nature of the project at hand, they will be frequently used. Also, as stated earlier, they should be designed in such a way that they will *never* abort the calling program (even for the most careless user).

In the area of programming to load data into the base, internal checking and total freedom from errors must be the guide posts. Although by this time the data will have had checks applied, it is often not practical, from a cost standpoint, to precheck every field of every data item. In the case at hand, this was particularly true of the call detail. It was necessary to extract the 56,000 calls desired from several million supplied (the received data filled 27 magnetic tapes). Since a single pass over this part of the data cost nearly seven thousand dollars (and was also a difficult job from an operational viewpoint), it was clearly desirable that the required program would not have to be repeatedly rerun due to program bugs. To this end, elaborate pretesting was done, and the input tapes were processed 3 or 4 at a time with extremely detailed summaries of counts, errors, *etc.*, printed. The printing of such summaries could be considered normal practice; however the crucial issue is that the output of each run be thoroughly scrutinized before submitting the next run. This point is often overlooked in the naive belief that, if a job terminated normally, it produced the desired results.

Another facet of loading data onto the final base is the requirement for some form of bookkeeping. Since we are dealing with thousands of data sets stored on a random access medium, it is clearly desirable to know the location of each. The ease with which this may be accomplished is, in part, a function of the random access software used. The software used in this base made the job easy (except for the necessity of maintaining literally hundreds of pages of records).

Finally, the base must be checked using the read routines, and the data base must be protected. Presumably, the random file would be backed up on tape. Unless the cost is prohibitive, two backup tapes should be considered a minimum — tapes can be destroyed (and if it can happen, it probably will).

4.7 The Data Base as Seen by the User

This and subsequent sections of this chapter contain increasingly machine-dependent information. Some slight knowledge of the Honeywell 6000 and Fortran will be assumed.

The users of this data base are provided with lists of the items contained in the base with corresponding data set names and numbers, three subroutines to read the data, and two "select" files containing job control cards for inclusion in the deck setup.

The use of lists of data set names (as opposed to files or some form of interactive query system) may surprise some readers. In fact, these offer several advantages. First, if properly set up, they are easy to use. Second, they are inexpensive to produce and maintain. Third, one need not have access to the computer to obtain required data names, and hence one may program without concern as to whether the computer is currently available. Finally, lists provide almost absolute security of the data. (The author accepts as axiomatic the fact that any data stored within a computer system can be read given a sufficiently competent programmer and sufficient time. In light of this, most present forms of "security" such as passwords, individual read permissions, *etc.*, are effective only in making things slightly more difficult for the unsophisticated user.)

The three read routines, accessible from a random library, consist of two for reading a single measure for all sample customers of a given class (respondents and non-respondents), and one for reading call detail for a single customer. The first two, NDRD and IDRD, are identical save for the method of communicating to the system which data set is required. NDRD uses mnemonic names as identifiers, IDRD uses integers.

The routine used to read the 1973 call detail for an individual customer is TPRD73. The calling sequence is similar to the previous routines except that the data desired is requested for a specified telephone NUMBER.

NUMBER admits to some slight control. Several of the sample customers subscribe to so-called "family plan" service under which two lines (with different numbers) are provided. The main number, generally published, is referred to as the Pilot number. The second, usually unpublished, is referred to as an Aux (auxiliary) line. The numbers need not be, and usually are not, consecutive.

Assume we have a subscriber with Pilot number 5551234 and Aux line 5556789. Requesting data on 5551234 (*i.e.*, passing this integer as NUMBER) will return detail of all calls on both lines. If the number is prefixed with 1 (*i.e.*, 15551234), detail of calls on the Pilot number only are returned. Prefixing a 2 (*i.e.*, 25551234) will return detail of calls made only on the Aux line (5556789). Equivalently, one could directly ask for data on 5556789. As previously mentioned, data sets exist on the base containing telephone numbers of sample customers, and additional sets exist wherein those telephone numbers for which data are missing are replaced by −1. Sets containing Pilot numbers and Aux line numbers of subscribers with "family plan" are also available.

The following eleven data items may be requested.

> Duration in minutes.
> Julian date.
> Time of the form HHMM (0000 to 2359) (H = hour, M = minute).
> Day of the week
>> 0 = Saturday
>> −1 = Sunday
>> 1 to 5 = Monday to Friday
>
> Terminating telephone prefix.
> Distance in yards between the "centers" of the originating and terminating Central Office Areas.
> Indication that the Central Office Areas are contiguous (1), non-contiguous (0), or the same (−1).
> Two-digit code for Central Office Area of terminating number.
> Two-digit code for zone of terminating prefix (a zone is an area, defined by the tariff, which typically consists of two Central Office Areas).
> Indication that the originating and terminating zones are

contiguous (1), non-contiguous (0), or the same (−1). Distance in yards between the centers of the originating and terminating zones as defined in the tariff.

In order to alleviate any potential problem of the user's not knowing the type of data returned (*i.e.*, INTEGER, REAL, LOGICAL, LITERAL, *etc.*) *all* data are stored and returned to the user as INTEGER.[5] Certain benefits in the area of computational speed and accuracy are also gained by this convention, but the main benefit is simplicity of use.

The above are absolutely all a user requires to use the data base. Of couse, a substantial number of other specialized routines have been made available to the users to accomplish frequently required analyses or summaries. Certain of these have been further generalized and are now available on the Bell Laboratories SCS Library. It should be noted, however, that these routines are truly ancillary and not a part of the data base system itself.

Next let us enter the world of reality, leaving the way the data base appears and examining the way it exists. The following section is, of necessity, highly machine-dependent.

4.8 Physical Description of the Data Base

The data base described resides on a 450 dismountable disc pack[6] and operates on the Honeywell 6000 computer at the Murray Hill, N.J., location of Bell Laboratories. The software used for actual random access is Jack L. Warner's Scatter Storage System, available on the SCS Library. It should be noted that the availability of this excellent package greatly facilitated the implementation of the base, and its exceptionally high quality in large measure underlies the base's reliability. We begin by examining some of the basic components.

The 450 pack is a random access storage medium of large capacity.[7] Approximately 40 million words of data may be stored on a single pack, and the data may be addressed on 64-word sector boundaries.

---

[5] This restriction causes certain seemingly strange conventions. For example, bills and their sub-components are stored as cents; distances are returned in yards (a compromise to the user community — the author would have much preferred to use meters, a far more rational measurement).
[6] Originally the data base was on a dismountable 190 disc pack; see Honeywell Information Systems, Inc., "DSS190 and DSS190B Disc Storage Subsystem Preference Manual, Series 6000", Publication DB37. The 450 has very similar characteristics except for a larger size.
[7] The space available is roughly equivalent to 11 full 9-track, 800 bpi tapes.

Multiple files may be defined within the pack, although that has not been done in this application. The 450 pack is very economical in terms of channel usage (I/O cost). Typically, equivalent data transfer to magnetic tape will be 8 times more expensive than to a 450 pack. The I/O charges to system permanent file are roughly comparable to the 450. The rental charge for a complete pack at Bell Laboratories is a small fraction of the cost of an equal amount of permanent disc storage.

The Honeywell system limits the use of dismountable packs to batch jobs only. However, auxiliary programs have been provided to facilitate copying selected portions of the pack to permanent files which may then be used by interactive systems. The computation center provides backup service for disc pack users (pack copied to tape) which provides the required protection of the data base.

The following paragraphs are not intended as a description of the Scatter Storage System. Rather, they examine a few selected aspects of the system and how these influenced the data base design.

Scatter Storage uses a hash coding technique to convert the data set name to a table pointer and thus find the actual disc address of the data. While hash coding is basically a sound technique, it suffers from susceptibility to so-called "collisions" — that is, cases in which the algorithm that "hashes" (encodes) the data set names produces the same pointer value for two different data set names. When collisions occur (and these are automatically detected), an alternate method of hashing is used. The process is repeated until a unique representation of the data set name is obtained. Clearly if many iterations are required on each name, then efficiency is greatly reduced. It is therefore prudent to examine the algorithm with a view to selecting optimal names.

When a Scatter Storage file is initialized, a parameter used in computing the size of the address table is specified. This parameter is referred to as the "power of 2". The parameter determines the maximum number of data sets which may be stored, the probability of collision for a set of randomly chosen names when a given number of data sets are stored, and, most importantly for this discussion, the working of the hashing algorithm.

For purposes of this base, twelve was the selected power of two, hence 2 to the 12th power (4096) data sets may be stored.[8] Since our data set names are all single-word variables, we examine this case only. Here the hashing algorithm divides the word containing the data set

---

[8] The mechanism used to allow storing more than 4096 sets will be discussed shortly.

name into 12-bit sections, bits 0 to 11, 12 to 23, and 24 to 35, right justifies each, sums the three numbers, and adds 1. Clearly if the data set names are integers in the range 0 to 4095, bits 0 to 23 will always contain zero, and the names will uniquely hash into their original value plus one. Collisions are thus impossible. This is the underlying strategy used in assigning set names for the majority of data sets in the base. These integer names under which the sets are stored are also the values used in calls to the read routine IDRD.

From the above it is obvious that some mechanism for associating the mnemonic names used by NDRD with the integer values is needed. The data divide themselves into natural groups: questionnaire data, bill data, *etc*. In construction of the names, the group was used to determine the last letter of the name. For example, names of all data sets obtained from questionnaires end in Q. Hence one finds that the data set containing the *sex* of the *head* of *household* is SEXHHQ. NDRD converts the data set name to the name of another data set containing all names of sets within the group. This is done by replacing the first letter of the name with a $, the next four with zeros, and leaving the final letter unaltered. NDRD then reads the list of data set names (assuming it is not already in core) and scans the set for the desired name. The position, together with an offset value stored at the beginning of the set, determines the integer name of the desired vector. The set is then read.

All data sets except those containing individual call detail are treated in this manner.

In the case of call detail, potentially almost 8,000 data sets exist. (The actual number is less due to lost data.) While methods of table lookup for assigning unique integer names to each is possible, it was judged to be too slow and to require too large a table. Instead, a new facility of Scatter Storage was utilized. By use of the routine DHADDR, multiple scatter storage areas, each with its own tables, may be established in a single file. Such an area was defined for each measure (date, time, zone, *etc.*). The integer indicator of measure is, in fact, a pointer to the appropriate area.

To allow access of data on pilot numbers only, the data for subscribers with family plan are stored twice — once for calls on pilot and Aux line, and again as data for the pilot number only.

The final test of any strategy, such as the ones described, is actual use. Experience has shown that any data set may be retrieved from the base for less than a half cent under normal service grade, or a quarter cent under economy grade. Although further reduction of cost is surely possible, the author, at least, is satisfied with the cost efficiency obtained.

4.9 Additional Data

As is too often the case, the exact amount of data which would ultimately be collected was not known in advance of the design of the data base. Billing data for several more months (billing periods beginning in February, March, and April of 1974) were obtained and added to the data base in exactly the same manner as described. Call detail for billing periods beginning in April, 1974, was also obtained. However, unlike the previous detail, both local and suburban data were available. Only a slight departure from the original design was required to accommodate the change. A routine named TPRD74 was written which appeared to users to be identical to TPRD73 previously described (except, of course, that 1974 data were returned). In order to access the detail on suburban calls, the users were instructed to prefix a "T" (for "timed") to the mnemonics used to retrieve local data. (For example, "CM" would return conversation minutes for local calls while "TCM" would return that data for suburban calls.) For those who preferred numeric codes, the rule was to add 100 to what the code was for the local data. These new data occupied about 12,000 additional data sets distributed in new scatter storage areas of the data base.

Later still, toll call detail for April, 1974, became available. For each toll call, the following information was provided:

>   date of call
>   terminating area code
>   originating area code
>   duration of call
>   class of call (station day, person day, *etc.*)
>   type of call (credit card, collect, direct dial, *etc.*)
>   time of day
>   charge.

Due to the anticipated use of these data, it was deemed advisable to store them, with selected demographic variables, on a separate data base maintained on a permanent disk file. This base, implemented by Susan J. Devlin and I. Lester Patterson, was essentially identical in design to the main data base. Since the quantity of data was much smaller, segmentation of data areas was found unnecessary (however this fact was transparent to users). Routines similar to NDRD and IDRD were used to access the demographics, while a routine similar to

TPRD74 was used to access the call detail. Due to the differences in the toll and local detail obtained, some different mnemonics were used (*e.g.*, "TYPE" for type of call).

4.10 With Perfect 20/20 Hindsight

The data base described has now been in use for about six years. It would seem appropriate to look briefly at certain problems that occurred and how the job might have been done better. Note that no criticism is implied in this section, but it has wisely been said that a man who makes a mistake and does not learn from it makes another mistake. A reasonable starting point might be the choice of the form of the mnemonics used by NDRD, the routine used to access demographic and billing data. For reasons of both efficiency and user convenience, names used were 6 characters (one word) long. The first 5 characters identify the contents of the data set; the last character identifies the source of the data. As mentioned, names ending in "Q" identified data obtained from the questionnaire. When billing data were initially placed on the base, two months were available, and names were generated ending in "O", and "A". The rather obvious meanings were October and April. In each set of names, the first 5 characters were repeated. For example the name "TOLLAx" referred to the toll charges on bills of questionnaire respondents for month x. As more data for other months were added, the mnemonic significance of the last character was unfortunately lost since data were available for both April, 1973, and April, 1974. Clearly both sets of names could not end with "A". It seemed preferable, however, to retain the original form rather than to restructure the entire data base, particularly since users would have to learn the new protocols, and existing programs would be difficult to keep working. Surely a better design, had the problem been known in advance, would have been to use longer names. One reasonable form would be "TOLLA.APR74" — toll for respondents for April, 1974, easily differentiated from "TOLLA.APR73". In fact, a base similar to the one described but using this form of longer name has been generated for use in another project.

The second problem one should address would seem to have no simple solution in terms of data base design. In the narrow view of the base, it might, in fact, be argued that there is no problem. More broadly, however, we are faced with the fact that data, like people, grow old. Since the questionnaire was administered in 1973, certain fields require special attention when used with data collected in 1974. An obvious example is the number of years a customer had lived in Chicago. Those who had responded "less than one year" in 1973 had, in fact, moved to the "1 to 2 year" category by 1974. While the "fix" needed here is rather obvious, it is less clear what should be done to groups spanning several years. Worse still are questions regarding what one should do to measures such as income. Fortunately, a number of other fields such as race and sex can be considered time invariant. These matters, while troublesome, are really not problems in the data, but rather problems for the data analyst. As such, no attempt was made to change the data base. Correction schemes used in the various analyses are discussed in other chapters. Also worthy of note is the problem of lines which were disconnected or reassigned. The first case presents no real problem in terms of the base as no data will exist and that information will be reflected. Reassigned numbers present more of a problem. Demographic information will surely be wrong, since we are dealing with a different customer — potentially even a business. However, the bill and calling data are correct, and could possibly be thought of as a form of replacement (although doubtless a very poor one). The decision was made to include data on the reassigned lines in the base, and make available files of the affected lines to users. Again the decision as to how (or whether) these data should be used was left to the analyst; no analyses reported in this book used the data for reassigned lines. Since the data were, in fact, accurate and could easily be identified and discarded if desired, it appears that this was the correct procedure.

4.11 Conclusions

Design and implementation of data bases is a challenging area. While software literature abounds with designs for "the ultimate all-pupose data base structure" (most of which differ widely), it would appear that none exists. Whether one *ever* will emerge seems doubtful. What does exist is a body of strategies and methods which may be employed to tailor a base to the data at hand in a manner similar to that described here.

The base described here appears to be successful from the standpoints of ease, economy of use and reliability. This is satisfying.

Appendix 4.A

Random and Sequential Data Access

In broad terms, a data base may be structured in one of two ways; sequential access or random access. In a sequential base, the records are necessarily retrieved (but not necessarily processed) in a fixed, predefined order. That is, in order to reach the nth record, one must read the preceding $n - 1$ records. Typical examples of this form of structure are decks of data cards and tape files. A certain degree of flexibility in reading records can be obtained by use of rewind and backspace commands. (Note that these commands work for cards since, during program execution, it is not the cards, but a file containing images of the cards, that is being read.) If only minor deviations from sequential processing are required, this form of data structure may serve well. Certainly it possesses two significant advantages. First, it is easily understood by anyone with even a limited knowledge of programming. Second, sequential files may be handled directly by almost all high-level languages.

The second major type of structure is the random access base. Here one specifies (in some fashion) the record to be read, and it is read directly; that is, read without the necessity of reading any other record. Of necessity, the base must reside on a file medium capable of random access. Normally, this would be a magnetic disc or drum. (It should be noted that discs and drums are, by their physical nature, random access devices. When used for sequential files, many systems cause these units to simulate magnetic tapes — inherently a sequential medium.) The efficiency, if not the convenience, of this form of structure is of course apparent. Further a random base may be used sequentially by simply reading the records in order; $1, 2, 3, ..., n$. Why, then, are random access bases not *always* used? Fundamentally, there are three reasons. First, many computer users have no understanding (and often a fear) of random access files. Second, several high level languages (such as the current standard Fortran) include no capability for reading such files. Additional routines (normally written in assembly language) must be used. The third reason, and in some sense the least valid, is cost. Charging algorithms used on most systems apply a far higher cost to disc, drum and disc pack storage than to magnetic tapes. Punched cards, in many cases, are essentially free. The point that is overlooked is that the cost of reading disc or drum is normally considerably lower than that of reading tape, and immensely lower than reading cards. The cost of the programmer's time must also be considered. Analyses which may be coded in a simple, straightforward way using a random base may require considerably more effort to code for a base which is sequential. This is particularly true if any attention is paid to program efficiency. (One may, of course, argue that the time required to generate a random base is greater than that required for one which is sequential. Experience shows that this is not necessarily true given a reasonably competent programmer and adequate software.) One must, then, conclude that, unless a base will *only* be used in a sequential fashion, random access should be considered (and probably selected).

CHAPTER 5

Box Plot Analysis of Local and Suburban Calling Frequencies and Conversation Times

*Belinda B. Brandon, Elsa M. Ancmon,
Carole C. Finck, and Robert H. Groff* [1]

5.1 Introduction

The present chapter analyzes the relationship of local and suburban usage with certain demographic household characteristics by means of the "notched box plot" display technique. The displays are supplemented by rank sum tests for significant differences in medians. The particular demographic characteristics that are chosen for presentation are race, income, and the age of the head of household. The analysis treats each characteristic separately.

For many of the results that are displayed below, there are two themes that the reader might keep in mind. One is that, since local and suburban calls have prices attached to them, one should expect to observe high-income households making more calls than low-income

---

[1] Carole C. Finck and Robert H. Groff were responsible for the programming during the formative stages of this work. Elsa M. Ancmon programmed the extensions that constitute the results presented in this chapter. Paul S. Brandon, who is not in the list of authors, served as a consultant during the analysis that led to this report.

households do, other things equal. We shall refer to this as a positive income elasticity effect. The second is that people have communities of interest. It seems likely that people will tend to call other people who have similar demographic characteristics to their own. For example, people may tend more to call other people of like income and like race. Chicago's suburbs have a higher proportion of people who are white and who have high incomes than Chicago itself does. So one might expect that white and high-income customers in Chicago should have a higher proportion of their calls going to the suburbs than black and low-income customers have. The relevant results in this chapter (and in Chapters 6 through 8) are consistent with these expectations, although, strictly speaking, Chapter 6's multivariate analysis provides the appropriate test of these hypotheses.

The chapter is divided into eight sections. Following this introductory section, the data are discussed. A tutorial discussion of the notched box plot method of presentation comprises the third section. The fourth reports the results of the analysis of Chicago local residence calls, including customers' numbers of calls, average durations, and total conversation times, as well as the results of a comparison of actual measurements with sampled customers' estimates of the numbers and durations of their calls. The next portion of the chapter reports the results of the analysis of calls placed to the suburbs by the sampled customers: the numbers of such calls, average durations, total conversation times, and numbers of message units are discussed. The sixth section combines local and suburban calls into total message units and reports on their analysis. The seventh portion of the paper examines the customer's choice of message-unit allowance as related to the number of local calls, average duration of local calls, total local conversation time, and total message units. The final section is a concluding note.

5.2 The Data

Recall that demographic information on the sampled customers was obtained by questionnaire in 1973. This chapter utilizes the information on race, income, and age. Also, information was collected in 1974 on sampled customers' bills, from which this paper utilizes only the choice of message-unit allowance. This chapter also uses data on the number of local and of suburban calls made by each customer, the duration of each call, and the prefix of each called number. These billing and calling data were for billing periods that began in April, 1974.

5.3 Tutorial on the Notched Box Plot

The technical method of display in this chapter is the "notched box plot", which is a five-point graphical summary of the distribution of a set of observations.[2] The advantages of the box plot are that a great deal of information can be presented in an easily understood fashion; that its measure of central tendency — the median — is little affected by extreme points; and that it measures the dispersion of the data robustly. Refer to the box in Figure 5.B.1 in Appendix 5.B for an example. That box plot is calculated from the entire sample's observations of households' numbers of local calls. The range of the lowest quarter of these observations is symbolized in the box plot by the distance between the lower extreme (one call) and the bottom of the box (33 calls), which is the lower quartile. The second quarter of the data is represented by the interval between the bottom edge of the box and the line running through the box. This line delineates the median (69 calls). The third quarter of the observations is depicted by the distance from the median line to the top of the box (at 129 calls), the top quartile. The data's top quarter is represented by the line from the top of the box to the highest point on the graph, which in this example is indicated to be off the scale at a value of 638.

The vertical size of the indentations in the sides of each box — called "notches" — is an approximate confidence interval around the median. In this book, for any box plot for the entire sample, the notches around the median m are specified as

$$m \pm 1.96\, se,$$

where se is the standard error of the median. From large-sample theory for Normally distributed data, the standard error is estimated by

$$se = \frac{0.926 R}{\sqrt{n}},$$

where R is the distance between the quartiles and n is the number of observations.[3] Thus, the notches in this case are an approximate 95 percent confidence interval around the median; i.e., in successive random samples from some population, notches that are calculated in this

---

[2] The box plot without notches was originated by John W. Tukey, *Exploratory Data Analysis* (Reading, Mass.: Addison-Wesley, 1977), pp. 39-41. The notches are presented in Robert McGill, John W. Tukey, and Wayne A. Larsen, "Variations of Box Plots", *The American Statistician*, v. 32, no. 1 (February, 1978), pp. 12-16.

[3] McGill, Tukey, Larsen, *op. cit.*

manner should contain the true median of the population approximately 95 percent of the time.

Most of the box plots are used to compare the telephone usage of one demographic group with that of another. For such comparisons in this book, the notches around each median are

$$m \pm 1.7\ se.$$

The notches are used primarily to provide a rough visual impression as to whether a difference between two groups is statistically significant: if the notch for one box does not overlap the notch in another box from an independent sample, then one can reject the hypothesis that the samples corresponding to the one box and to the other box are drawn from the same population at about the 5 percent significance level. The precise technical statement is that if the two groups from which the samples are drawn really had the same median, then such a large difference in the two sample medians could occur by chance (through the randomness of the sample drawing) less than approximately 5 percent of the time. Figure 5.B.2 is an example of a significant difference; 5.B.3 displays insignificant differences. The multiplier of 1.7 is chosen as a compromise constant for every comparison whereas strictly speaking a different multiplier should be used for a comparison between each pair of groups, depending on relative standard errors of the medians. If two samples have near equal-sized standard errors of their medians, then the appropriate multiplier would be closer to 1.4; if their standard errors were very different, then the appropriate multiplier would be near two.[4]

These notches are useful because they aid the eye in judging whether a difference is meaningful or not, but, because they are approximations, the result of the Wilcoxon-Mann-Whitney rank sum test for a difference in medians is reported in Appendix 5.A for every pair of groups. That appendix also includes an explanation of the latter test. Whenever the conclusion of this test differs from that of the notches, the discrepancy is that the rank sum test finds a significant difference that the notches do not. (Typically such an event occurs when the estimated standard errors of the medians of two groups are nearly equal, so that the more appropriate notches would have been near 1.4 standard errors around the median instead of the 1.7 used in the figures.) For expositional convenience, statements in the text about statistical significance are consistently based on the rank-sum test.

---

[4] McGill, Tukey, Larsen, *op. cit.*

Since the theory underlying the calculation of the notches assumes that the data are drawn from the Normal distribution, in this chapter all of the calling data are transformed by adding the arbitrary constant 1/6, then by taking the logarithm. These transformed data still are not Normally distributed. But taking the logarithm reduces the departure from Normality for each of the calling measures. A visual impression of the closeness to Normality in one instance can be obtained by examining Exhibits 5.1A and 5.1B on the following page, which are quantile-quantile $(Q-Q)$ plots of the raw data and of the transformed data, respectively, for the entire sample's observations of numbers of calls.[5] If a data set were Normally distributed, then its $Q-Q$ plot would be a straight line with slope equal to the standard deviation of the sample, and intercept equal to the mean of the sample. Neither Exhibit 5.1A nor 5.1B matches that description, but the $Q-Q$ plot of the transformed data in 5.1B is closer to it than is 5.1A. Intuitively, the notches should be more accurate as derived from the transformed data. Fortunately, there is some evidence that the accuracy of the notches as confidence intervals is not very sensitive to departures from Normality.[6]

Once the confidence intervals for the transformed data are found, the end points of the intervals are retransformed into the raw scale so that they may be entered on the box plots of the raw data. Since a logarithmic transformation leaves orderings unchanged, an overlap or a lack of overlap of the notches of different groups is preserved as one applies a retransformation to the raw scale, although the spacing between notches should change.

Sometimes the variance of the data is so large relative to the number of observations that the notches fall outside the interquartile range. When this event occurs, the notches are not shown; instead, for aesthetic reasons the box plot is drawn with dashed lines.

5.4 Local Calls

This section analyzes local Chicago residence calls for the total sample and for selected demographic characteristics — race, 1972 family income, and age of the head of household. We relate these demographic characteristics to the number of local calls, average duration of

---

[5] Martin B. Wilk and Ramanathan Gnanadesikan, "Probability Plotting Methods for the Analysis of Data", *Biometrika,* no. 1, v. 55 (March, 1968), pp. 1-17.
[6] McGill, Tukey, Larsen, *op. cit.*

80 *The Effect of Demographics on Telephone Usage*

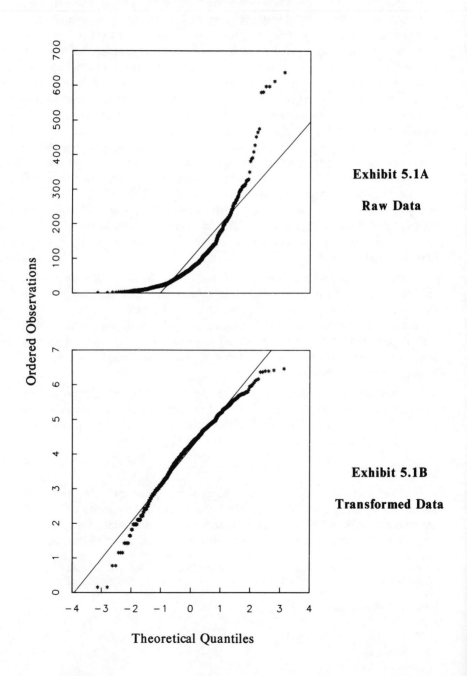

Exhibit 5.1A

Raw Data

Exhibit 5.1B

Transformed Data

local calls, and total local conversation time. We also look at the relationship of the number and duration of calls estimated by each customer with observed values. In the following discussion of results, the convention is typically applied that a difference between two groups is called "statistically significant" if it is significant at the 5 percent level or better.

5.4.1 Numbers of Local Calls

Figures 5.B.1 through 5.B.4 portray on their vertical axes the summary statistics of the distribution of sampled households' monthly numbers of outgoing local calls; Figure 5.B.1 does so for the entire sample, and 5.B.2 through 5.B.4 do so for the racial, income, and age groups, respectively, on the horizontal axes. In this chapter, results for only the two principal racial groups are presented — white and black customers; their sample sizes are 232 and 146, respectively. Twenty-two Spanish-speaking, 13 "other", and 7 "no answers" are omitted. Several income groups are aggregated into three — less than $5,000, at least $5,000 but less than $15,000, and at least $15,000; the group, comprising 16 percent of the sample, that did not provide income information is omitted. All the information on age of the head of household is used, although again no results are reported for the 15 customers who declined to divulge their age.

Figure 5.B.1 has already been discussed above as an example of the box plot. Recall that it shows that the median number of local calls for the entire sample is 69; that the approximate 95 percent confidence interval around the median is 62 to 77 calls; that the middle half of the sample data lies between 33 and 129 calls; and that the entire range of the sample data is one to 638 calls. Note that the sample size is shown below the box plot to be 573 customers.[7]

In Figure 5.B.2, one finds that the median number of local calls for white customers (61.5) is less than that for black customers (115.0).

---

[7] It should be pointed out that — even though the size of the entire sample for this chapter is 573 — for any analysis involving demographics, the maximum possible sample size is 420. The difference of 153 is the number of customers in the sample for whom we have no information from the questionnaire. Since the latter group may have different telephone usage characteristics from the rest, box plots for the entire sample are not directly comparable to those for demographic groups. Sample sizes for particular analyses may fall below 420 because of missing billing data, customers' refusals to answer particular questions in the questionnaire, no calls for a customer with which to calculate average durations, or, in the case of race, omissions of some groups. The minimum sample size for any display is 294.

In fact, the median number of calls for black customers exceeds the top quartile for white customers. The difference in medians is statistically significant at the 5 percent confidence level. A notable finding in this and subsequent figures is that the variation in usage among customers within a demographic group is large relative to differences between groups.

Figure 5.B.3 analyzes usage by the three income groups. The median number of calls rises monotonically with income, although none of the differences is statistically significant.

Figure 5.B.4 displays the relationship between the number of calls and age of the head of household. The medians of the four groups under 55 years of age are not significantly different from one another, nor are the two oldest age groups significantly different. But both older groups are significantly lower than every younger group. The pattern for the age of the head of household presumably is affected by, but not completely explained by, the relationship between age of the head of household and household size. The median numbers of members for households with heads of each of the age groups are shown below:

Age of Head	Median Number in Household
Under 25	2
25 to 34	3
35 to 44	4
45 to 54	3
55 to 64	2
65 or over	2

It may be worth noting that we experimented with some interactions between demographic characteristics. For example, we generated box plots of the number of local calls by income for each racial group. While the calling rate for black customers is higher than that for white customers for each level of income, the differences in the calling rate between income levels is similar for the two races. No striking race-income interactions were found for other calling measures; thus, no such plots are displayed.

5.4.2 Removal of Rounding Bias from Durations

Section 5.4.3 below analyzes the average durations of households' local calls. But, before discussing the results, we describe an adjustment to the data. The data as received are biased by the following rounding process: a call that has an actual duration (after the called party lifts the receiver) of at least 0.1 minute[8] and less than 1.1 minute is recorded by the equipment in Chicago as having had a duration of 1.0 minute. Similarly, at least 1.1 and less than 2.1 is recorded as 2.0 minutes, and so forth. We have attempted to estimate the resulting bias and remove it from the recorded durations. The non-technical reader may want to skip directly to Section 5.4.3.

We handle the durations as follows: Assume, as is frequently done, that the actual durations of local calls for a given customer during a particular hour are exponentially distributed. (Some analysis with $Q-Q$ plots supports this assumption.) The maximum likelihood estimate of the mean of the actual durations for the customer is obtained as follows:[9] The probability of a recorded duration of t_i for call i given the mean μ of actual durations is

$$Pr(t_i|\mu) = \int_{t_i-0.9}^{t_i+0.1} \frac{1}{\mu-0.1} \exp\left(-\left[\frac{x-0.1}{\mu-0.1}\right]\right) dx, \quad (5.1)$$

so

$$Pr(t_i|\mu) = \left[\exp\left[\frac{1}{\mu-0.1}\right] - 1\right] \exp\left(-\left[\frac{t_i}{\mu-0.1}\right]\right). \quad (5.2)$$

The probability of recording a set of n durations is

$$Pr(t_1,\ldots,t_n|\mu) = \left[\exp\left[\frac{1}{\mu-0.1}\right] - 1\right]^n \prod_{i=1}^{n} \exp\left(-\left[\frac{t_i}{\mu-0.1}\right]\right). \quad (5.3)$$

The log-likelihood is

$$\ln Pr(t_1,\ldots,t_n|\mu) = n\left[\ln\left[\exp\left[\frac{1}{\mu-0.1}\right] - 1\right] - \frac{\bar{t}}{\mu-0.1}\right], \quad (5.4)$$

where $\bar{t}$ is the arithmetic mean of recorded durations. Note that the rounding rule results in $\bar{t} \geq 1$. The log-likelihood is maximized by selecting the following as an estimate of μ:

---

[8] In the areas served by the North Center Revenue Accounting Office, a call of actual duration less than 0.1 minute is not recorded as having occurred. In the Marquette Park Office, a call of duration greater than zero but less than 0.1 minute is recorded as 1.0 minute in our data base. But from another data set it is known that there are few such calls, so their possibility is ignored.

[9] A maximum likelihood estimator of distribution parameters in a similar context was also used independently by Susan J. Devlin and by Aya Cohen in unpublished work at Bell Laboratories.

$$\hat{\mu} \begin{cases} = 0.1 - \dfrac{1}{\ln\left(1 - \dfrac{1}{\bar{t}}\right)} & \text{if } \bar{t} > 1 \\ = 0.1 & \text{if } \bar{t} = 1. \end{cases} \quad (5.5)$$

This estimate is consistent; that is, if durations for a customer are truly exponentially distributed, then, as the number of calls gets large, $\hat{\mu}$ approaches the true expected actual duration. Based upon the asymptotic approximation and the exponential assumption, an estimate of the degree of bias in terms of the mean of recorded durations is

$$\text{Bias} = \bar{t} - 0.1 + \dfrac{1}{\ln\left(1 - \dfrac{1}{\bar{t}}\right)}. \quad (5.6)$$

This bias is estimated for each individual for each "hour" (where one "hour" is, for example, all the periods between 4 and 5 p.m. on all weekdays during the month of study). The estimated bias is subtracted from the duration of each of his calls during that hour. (Given the assumption of the exponential distribution, the bias is the same for every call, regardless of duration.) It is this asymptotically unbiased data set that is used for subsequent calculations. In order to give the reader an impression of the size of this adjustment, a few illustrative recorded average durations are shown below together with the corresponding estimated "actual" average durations and asymptotic bias in the recorded durations:

Recorded Average Duration	Estimated "Actual" Average Duration	Asymptotic Bias
1.0	0.100	0.900
1.1	0.517	0.583
1.5	1.010	0.490
2.0	1.543	0.457
3.0	2.566	0.434
4.0	3.576	0.424
5.0	4.581	0.419
10.0	9.591	0.409
∞	∞	0.400

The reader's intuition regarding this bias may be aided by explaining how it is that 0.4 minutes is the limiting value of the bias. As the mean duration gets large, the exponential distribution approaches the uniform distribution for any interval. A call of actual duration anywhere between $D - 0.9$ and $D + 0.1$ minutes gets recorded as D minutes. If durations are uniformly distributed within that interval, then the expected actual duration for a call that is recorded as D minutes is

$$\int_{D-0.9}^{D+0.1} x \, dx = D - 0.4,$$

so the bias is 0.4 minutes.

It may be worth noting that this adjustment for bias provides a data set that is more useful for distinguishing calling patterns of groups and for evaluating differential loads on the telephone network, as compared to the original data set. But it is the original data set that is more useful for estimating potential revenues from suggested pricing changes, because any new price structure in Chicago may very well use Illinois Bell's existing rounding convention.

5.4.3 Average Durations of Local Calls

Figure 5.B.5 summarizes the distribution of the households' asymptotically unbiased average durations of local calls on the vertical axis for the whole sample; Figures 5.B.6 through 5.B.8 plot it against demographic groups on the horizontal axes. The median of all sampled customers' average durations of local calls, as shown in Figure 5.B.5, is 4.9 minutes. When one compares racial groups, which are shown in Figure 5.B.6, one can see that black customers have a median average duration that is significantly higher than that for white customers; the difference is 1.3 minutes. Figure 5.B.7 shows that the median of the household's average durations falls as income rises. The differences between the under-$5,000 group and both of the other groups are statistically significant. Figure 5.B.8 shows no systematic relationship between age of the head of household and the medians of average durations.

5.4.4 Total Local Conversation Times

Figures 5.B.9 through 5.B.12 display the month's total local conversation times on their vertical axes; the first of these figures is for the entire sample, and the others have demographic groups on their horizontal axes. The conversation times again have their estimated biases removed. Since black customers have a higher median number of local calls as well as a greater median of their average durations, one might expect that their median total local conversation time would be

86 *The Effect of Demographics on Telephone Usage*

greater than that of white customers. In fact, as shown in Figure 5.B.10, the median total local conversation time for black customers is over twice that of white customers, and the difference is statistically significant. There is no systematic relationship between total local conversation time and income group (see Figure 5.B.11). Figure 5.B.12 shows that the two older groups (55 to 64 and 65 or over) are significantly different from the younger groups except for a near miss in the case of the under-25 group versus the 65-or-over group. However, neither the set of younger groups nor the set of older groups is significantly different within the set.

5.4.5 Customer-Estimated vs. Observed Numbers and Durations of Local Calls

In order to understand how a customer will react to a price change, it may be helpful to determine the customer's perception of his present telephone usage. Figure 5.B.13 illustrates that the recorded number of local calls tends to fall short of customers' estimates. On the vertical axis is the observed number of calls per day and on the horizontal axis is the customer-estimated number of calls per day. Here the notches are approximate 95 percent confidence intervals around the medians. The dots on the diagonal represent the set of points for which the observed number of calls would equal the customer-estimated number of calls. In general, the median number of calls actually recorded rises as the number estimated by the customers rises. But many customers err: for a majority of the customers who estimated one or two calls on the 1973 questionnaire, the number of calls actually recorded in April, 1974, exceed their estimates.[10] By contrast, for a majority of the customers who estimated more than two calls, their usage falls short of their estimates; and with an increasing number of estimated calls, the difference tends to become greater. For instance, at three estimated calls per day the median observed number of calls is 2.4 (a difference of 0.6 calls), but at six claimed calls per day the median is 3.2 (a difference of 2.8 calls).

Consider a different methodology. One might want to calculate the mean number of local calls for the entire sample and compare it to the mean estimate. A difficulty is that one does not know how to represent those who estimate "9+" calls per day. But regardless of what number one uses for 9+, one comes to the same qualitative

---

[10] According to the 1973 calling data, a customer's estimate also tended to be too small if he reported one call, but not if he reported two calls. The results for 1973 and 1974 are otherwise qualitatively similar.

conclusion: the sample of customers overestimate their number of local calls. For example, if one assumes that the mean estimate of those who said 9+ calls was 12 calls, then the mean estimate for the entire sample in April, 1974, was 4.05 calls, which is significantly greater than the mean recorded number of calls, 3.51 (the z-statistic is 2.35).[11]

The customer may be overestimating his usage for one or a combination of the following reasons, besides the possibility that he simply estimates poorly. He may think that the definition of a local call includes calls to the suburbs. He may be including all attempted calls rather than just completed calls. Finally, he may be counting many incoming calls in addition to outgoing calls.

Referring to Figure 5.B.14, one can see that the median recorded average duration of local calls (with bias removed) rises with the customer-estimated duration. But a tendency to overestimate is evident, especially in the cases of those who reported a 5-to-10-minute or over-10-minute average: the recorded median for those who report a 5-to-10-minute average is 4.8, so over half of them are recorded as having an average less than that range; and over 75 percent of those who estimated an average of over 10 minutes actually have a recorded average less than 10 minutes. (There is no analysis of mean estimates because it would be so sensitive to the choice of how to represent "less than 5 minutes" and "greater than 10 minutes".)

5.5 Suburban Calls

Recall that a suburban call is defined for the purpose of this book to be a completed call placed by a Chicago customer to a specified band of Illinois suburbs. While a customer incurs one message unit for a local call, regardless of its duration, the customer is charged according to the duration of suburban calls. The initial charges for these suburban calls vary from 1 to 8 message units, as determined roughly by the distance of the call.

This section consists of four parts that deal with, respectively, customers' numbers of suburban calls, average durations of suburban calls, total suburban conversation times, and total suburban message units. Each part reports on four figures, one of which aggregates all sampled customers together, and three of which disaggregate them into

---

[11] The results using April, 1973, data are similar. The mean estimate, taking 9+ to be 12, is 4.08, which is significantly greater than the mean recorded number of calls, 3.44 (the z-statistic is 3.40).

demographic groups by race, by income, and by age of the head of household.

5.5.1 Numbers of Suburban Calls

Figures 5.B.15 through 5.B.18 deal with the monthly number of suburban calls for each sampled customer. As shown in Figure 5.B.15, the median number of suburban calls for the entire sample is 5. If one were to break down suburban calls by initial message unit charge, one would find that the greatest number of these calls have a three-message-unit initial charge. The median number of calls is zero for all other initial message unit charges. For this reason we do not show plots by initial message charge.

Figure 5.B.16 graphs numbers of suburban calls on the vertical axis and racial groups on the horizontal axis. The median number of suburban calls for white customers is 9, but for black customers is only 2; this difference is statistically significant.

Figure 5.B.17 depicts the relationship between the number of suburban calls and income groups. As income increases, the median number of suburban calls increases monotonically. The median number of suburban calls for customers whose income is under $5,000 is 2. But for customers whose income is $15,000 or over, the median has risen to 11 calls. The median of each income group is significantly different from each of the others.

Figure 5.B.18 shows that the median number of suburban calls perhaps tends to rise with age up to 7.5 calls for the 45-to-54 age group. It then declines to 3 calls for households that are headed by a person 65 or over. However, only the difference between the 25-to-34 and 65-or-over age groups is statistically significant. In the light of the multiple comparisons problem explained in Chapter 1, one may not want to emphasize this one difference.

5.5.2 Average Durations of Suburban Calls

Customers' average durations per suburban call, with estimated bias removed, are depicted by Figures 5.B.19 through 5.B.22. The median of the average durations of the customers in the entire sample is 4.6 minutes, as can be seen in Figure 5.B.19. (The "sample size" at the bottom of each box in this part on average durations of suburban calls is the number of customers who made at least one call. Since many customers made no suburban calls in the month of study, the sample sizes shown are smaller than for box plots of other calling measures.)

Figure 5.B.20 exhibits the relationship between average durations of suburban calls and racial groups. White customers have a median average duration for suburban calls of 5.5 minutes, which is more than twice the median average duration for black customers (2.4 minutes); the difference is statistically significant.

Figure 5.B.21 graphs average durations of suburban calls on the vertical axis and income groups on the horizontal axis. None of the differences is statistically significant.

Figure 5.B.22 depicts how average duration relates to the age of the head of household. The median of households' average durations of suburban calls holds fairly steady with age to the 45-to-54-year group, then jumps for the older two groups. But significant differences are found only between the 55-to-64-year group and the three middle-age groups — 25 to 34, 35 to 44, and 45 to 54.

5.5.3 Total Suburban Conversation Times

Figures 5.B.23 through 5.B.26 examine customers' monthly total suburban conversation times (with estimated bias removed). The median total suburban conversation time for all sampled customers is shown as 19.9 minutes in Figure 5.B.23.

Figure 5.B.24 shows that the median total suburban conversation time for white customers is 50.5 minutes and for black customers is only 2.7, a statistically significant difference.

In Figure 5.B.25 one sees that the higher is the income, the greater is the total suburban conversation time. The median total suburban conversation time progresses from 8.0 minutes for those with incomes under $5,000, to 26.0 minutes for the $5,000-to-$14,999 group, and to 54.6 minutes for customers whose income is $15,000 or over. Each group is significantly different from each of the other two.

The last figure on total suburban conversation time — 5.B.26 — reveals no systematic relationship between total suburban conversation time and age of the head of household.

5.5.4 Suburban Message Units

For local Chicago calls, the number of calls equals the number of message units; but for suburban calls this equivalence does not hold. Recall that in order to initiate a suburban call, one incurs a number of message units that rises roughly with distance; then one incurs additional message units depending on both the distance and the duration of the call.

90   *The Effect of Demographics on Telephone Usage*

Figures 5.B.27 through 5.B.30 deal with suburban message units. The median number of suburban message units is 22 for the entire sample (see Figure 5.B.27). When one examines Figures 5.B.28 through 5.B.30, one finds that the relationship between suburban message units and the demographic variables is qualitatively the same as the relationship between total suburban conversation time and those demographic variables. Therefore, detailed comments need not be made for these figures.

5.6 Total Local and Suburban Message Units

In this section we describe the association of total local and suburban message units with demographic characteristics, utilizing Figures 5.B.31 through 5.B.34. Total message units are obtained by adding the number of local calls to suburban message units. Recall that a charge on the customer's bill is made for each message unit in excess of his message unit allowance, and that the customer pays a higher basic service charge for a larger allowance. Figure 5.B.31 presents the distribution of customers' total numbers of message units for the entire sample. The median monthly number of message units is 111.0.

Figure 5.B.32 shows the number of message units on the vertical axis and racial groups on the horizontal axis. White customers' median number of message units is 111.0 and black customers' is 131.5. If one refers back to Figure 5.B.2 and Figure 5.B.28, one can readily see that the median number of local calls for black customers is much higher than for white customers, but that the median number of suburban message units for black customers is much lower than for white customers. When the number of local Chicago calls and the number of suburban message units are combined, as in Figure 5.B.32, the difference between the racial groups is not statistically significant. It may also be interesting to note that the sample *mean* of the total local and suburban message units is actually higher for white customers — 179 — than it is for black customers — 172 — although this difference is also not statistically significant.

According to Figure 5.B.33, the median number of message units increases as income increases; the differences between the highest-income group and the other two groups are statistically significant. Recall that Figure 5.B.3 displays a positive but insignificant association between income and the number of local calls, and that Figure 5.B.29 displays a positive and significant association between income and suburban message units.

The number of message units versus age of the head of household is graphed in Figure 5.B.34. As the age of the head of household increases, the median number of total message units decreases. The median of the 65-or-over group is significantly different from the four youngest groups, and the 55-to-64 group is significantly different from the three groups between 25 and 54 years. Note that Figure 5.B.4 exhibits similar results for the number of local calls, whereas Figure 5.B.30 shows no such systematic relationship for the number of suburban message units.

5.7 Usage by Message-Unit Allowance

This section of the chapter examines how the sampled customers' mesage-unit usage is associated with their chosen message-unit allowances. The rate structure in Chicago is such that a residence customer can choose among a set of message-unit allowances, paying a fixed amount per month for that allowance so long as his usage does not exceed the allowance; if his usage does exceed the allowance, then he must pay an additional amount for each message unit in excess of the allowance. The message-unit allowances available in 1974 were zero, 30 (two-party), 60 (two-party), 80, 140, 200, and Unlimited.

In Figure 5.B.35, the total number of local and suburban message units is graphed on the vertical axis, and message-unit allowances are on the horizontal axis. In addition, for each allowance the figure indicates the range of the number of message units for which that allowance is the lowest-cost choice. For simplicity, the assumption is made that customers attach no premium for desirability of single-party service relative to the two-party service offerings (with 30 and 60 message-unit allowances). Under this assumption and the rates prevailing in 1974, in order to minimize outlay a customer who knows how many message units he would incur in a month should have chosen an allowance according to the following schedule:

Message-Unit Allowance	Usage of Message Units
0	0 to 7
30 (two-party)	8 to 48
60 (two-party)	49 to 84
80	85 to 121
140	122 to 181
200	182 to 440
Unlimited	441 or more

Except for the zero-allowance choice, which has too small of a sample to be relevant to the discussion, in Figure 5.B.35 the median of customers' numbers of message units grows monotonically with message-unit allowance. And, except for the small sample of zero-allowances, each of the medians fall within the range of the numbers of message units for which that allowance is lowest-cost. But, at least for this one month, there is a large proportion of customers who would have saved themselves some money if they had chosen a different allowance. Of course, one should not expect every customer to have the "right" allowance every month. Among other reasons, a customer's usage during a month is not known by him ahead of time; and a significant amount of time is involved in informing himself, making a decision about a change in allowance, and communicating with the telephone company. What has not been investigated is what proportion of customers did not make the "right" choice of allowance considering their usage over a long span of time. It may be worth noting that, after the period of this study, two new single-party allowance choices were added in Chicago — 110 and 170 message units.

5.8 Summary and a Concluding Note

In summary, we find that black customers have higher local usage but lower suburban usage than do white customers. Income has a positive association primarily with suburban calling and total message units. Households with heads of age 55 years or more tend to have lower local usage and total message units than do other households.

In this chapter, we have examined calling characteristics as related to demographic characteristics. Throughout, the box-plot method of analysis has been utilized. This method allows one easily to see what patterns exist without being misled by extreme observations. However, this technique also makes it difficult for one to examine many variables simultaneously so as to tell whether each variable still has an effect while controlling for variations in the others. That kind of examination is achieved by employing multivariate analysis. Such an analysis is contained in the next chapter.

Appendix 5.A

Wilcoxon-Mann-Whitney Rank Sum Tests

As a supplement to the notches in the box plots, this appendix shows the results of an alternative test for differences between pairs of demographic groups. The one selected is the Wilcoxon-Mann-Whitney rank sum test, which is a test for differences in medians, as the notches are, but which is non-parametric and does not rely on the assumption of a particular probability distribution. The procedure is as follows: the data on some usage measure for two disjoint groups are combined into one data set and are assigned ranks according to the values of the usage measure. If there is a set of more than one household with the same value, then each of them in the set is assigned the average rank of the set. For any one of the groups, the statistic

$$z = \frac{k + 1/2 - m[m+n+1]/2}{[mn[m+n+1]/12]^{1/2}}$$

is calculated, where k is the sum of the ranks of the households in the smaller group, and m and n are the sizes of the smaller and larger of the two groups, respectively. Asymptotically, under the null hypothesis that both groups are drawn from the same population, z is Normally distributed with zero mean and with unitary variance.[1] If one has no prior knowledge about the relative usages of groups, then a two-tailed test is appropriate. Each comparison of two groups yields a value for z. From each z, the probability of observing a value less than or equal to that z is calculated and reported in the table on the following page. If one utilizes a 95 percent confidence level, then one can reject the null hypothesis if the reported probability is either no more than 2.5 percent or no less than 97.5 percent. Any such significant difference is shown in bold type. If the reported probability is less than 50 percent, then the smaller group, which is listed to the left, has the lower average rank; if greater than 50 percent, the larger group to the right is ranked lower.

---

[1] See, *e.g.*, B. W. Lindgren and G. W. McElrath, *Introduction to Probability and Statistics* (New York: Macmillan, 1966), p. 231.

Probability of Observing a Rank Sum No Larger than Observed for the Smaller Sample if Two Samples are Drawn from the Same Population

Pairs of Groups		Local			Suburban			Local and Suburban
		Number of Calls	Average Duration	Total Conv. Time	Number of Calls	Average Duration	Total Conv. Time	Total Message Units	Total Message Units
Smaller	Larger				Probabilities				
Black	White	**0.999**	**0.999**	**0.999**	**0.000**	**0.000**	**0.000**	**0.000**	0.949
<$5,000	$5,000-14,999	0.403	**0.986**	0.829	**0.001**	0.204	**0.003**	**0.012**	0.163
<$5,000	≥$15,000	0.129	**0.997**	0.818	**0.000**	0.444	**0.000**	**0.000**	**0.008**
≥$15,000	$5,000-14,999	0.888	0.116	0.555	**0.999**	0.204	**0.995**	**0.999**	**0.982**
<25	25-34	0.618	0.357	0.559	0.182	0.635	0.270	0.182	0.377
<25	35-44	0.359	0.440	0.289	0.423	0.554	0.450	0.364	0.497
<25	45-54	0.676	0.220	0.430	0.228	0.676	0.316	0.214	0.554
<25	55-64	**0.998**	0.173	**0.978**	0.281	0.112	0.222	0.173	0.949
<25	≥65	**0.994**	0.281	0.971	0.652	0.204	0.445	0.448	**0.976**
35-44	25-34	0.898	0.268	0.811	0.091	0.619	0.111	0.098	0.340
45-54	25-34	0.323	0.771	0.632	0.369	0.404	0.344	0.308	0.248
55-64	25-34	**0.000**	0.856	**0.001**	0.298	**0.989**	0.424	0.482	**0.001**
≥65	25-34	**0.000**	0.595	**0.002**	**0.024**	0.922	0.181	0.094	**0.000**
45-54	35-44	0.116	0.902	0.407	0.857	0.698	0.764	0.783	0.399
55-64	35-44	**0.000**	0.943	**0.000**	0.789	**0.984**	0.930	0.909	**0.003**
≥65	35-44	**0.000**	0.704	**0.000**	0.195	0.904	0.531	0.381	**0.001**
55-64	45-54	**0.000**	0.595	**0.003**	0.443	**0.997**	0.706	0.719	**0.014**
≥65	45-54	**0.004**	0.338	**0.005**	0.045	0.952	0.243	0.151	**0.006**
≥65	55-64	0.656	0.282	0.455	0.081	0.302	0.133	0.110	0.235

Appendix 5.B

Box Plots of Local and Suburban Usage by Demographic Group

Figure Number	Title	Page
5.B.1	Number of Local Calls for Entire Sample	97
5.B.2	Number of Local Calls by Race	98
5.B.3	Number of Local Calls by Income	99
5.B.4	Number of Local Calls by Age of Head of Household	100
5.B.5	Average Durations of Local Calls for Entire Sample	101
5.B.6	Average Durations of Local Calls by Race	102
5.B.7	Average Durations of Local Calls by Income	103
5.B.8	Average Durations of Local Calls by Age of Head of Household	104
5.B.9	Total Local Conversation Time for Entire Sample	105
5.B.10	Total Local Conversation Time by Race	106
5.B.11	Total Local Conversation Time by Income	107
5.B.12	Total Local Conversation Time by Age of Head of Household	108
5.B.13	Observed vs. Customer Estimated Number of Local Calls	109
5.B.14	Observed vs. Customer Estimated Average Duration of Local Calls	110
5.B.15	Number of Suburban Calls for Entire Sample	111
5.B.16	Number of Suburban Calls by Race	112
5.B.17	Number of Suburban Calls by Income	113
5.B.18	Number of Suburban Calls by Age of Head of Household	114
5.B.19	Average Durations of Suburban Calls for Entire Sample	115
5.B.20	Average Durations of Suburban Calls by Race	116
5.B.21	Average Durations of Suburban Calls by Income	117
5.B.22	Average Durations of Suburban Calls by Age of Head of Household	118
5.B.23	Total Suburban Conversation Time for Entire Sample	119
5.B.24	Total Suburban Conversation Time by Race	120
5.B.25	Total Suburban Conversation Time by Income	121

5.B.26	Total Suburban Conversation Time by Age of Head of Household	122
5.B.27	Suburban Message Units for Entire Sample	123
5.B.28	Suburban Message Units by Race	124
5.B.29	Suburban Message Units by Income	125
5.B.30	Suburban Message Units by Age of Head of Household	126
5.B.31	Total Local and Suburban Message Units for Entire Sample	127
5.B.32	Total Local and Suburban Message Units by Race	128
5.B.33	Total Local and Suburban Message Units by Income	129
5.B.34	Total Local and Suburban Message Units by Age of Head of Household	130
5.B.35	Total Local and Suburban Message Units by Message Unit Allowance for Entire Sample	131

Figure 5.B.1

Number of Local Calls for Entire Sample

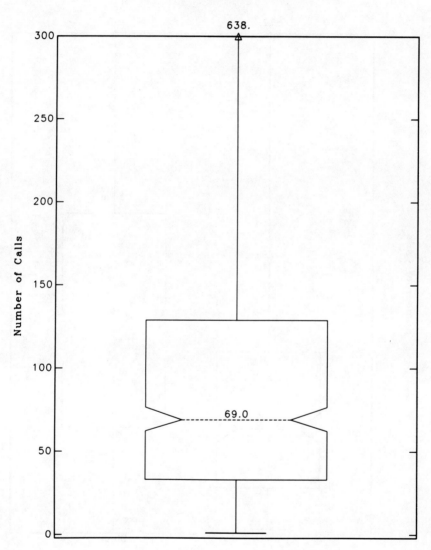

Sample Size 573

98 *The Effect of Demographics on Telephone Usage*

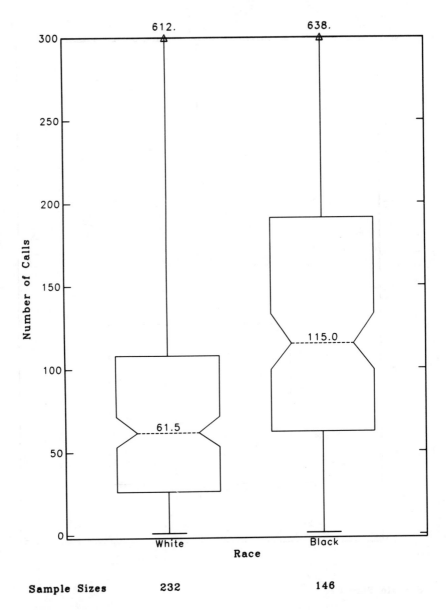

Figure 5.B.2

Number of Local Calls by Race

Figure 5.B.3

Number of Local Calls by Income

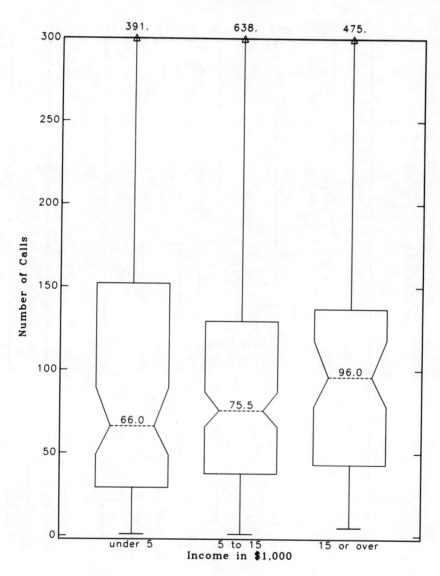

100 *The Effect of Demographics on Telephone Usage*

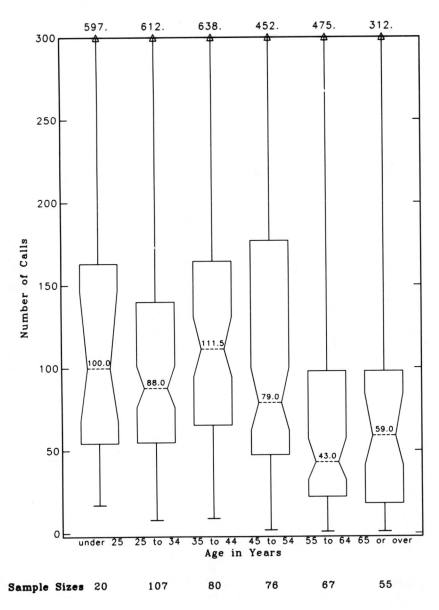

Figure 5.B.4

Number of Local Calls
by Age of Head of Household

Box Plot Analysis of Local and Suburban Calling 101

Figure 5.B.5

Average Durations of Local Calls for Entire Sample

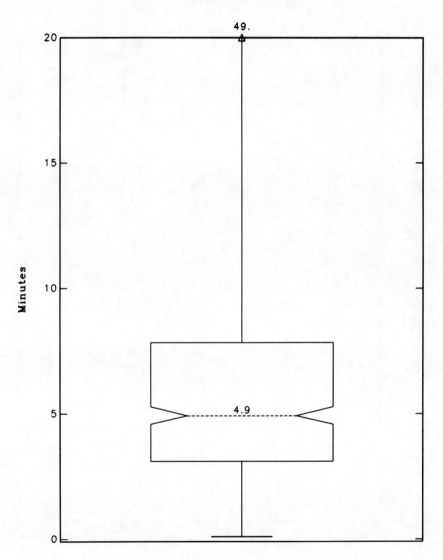

Sample Size 573

102 *The Effect of Demographics on Telephone Usage*

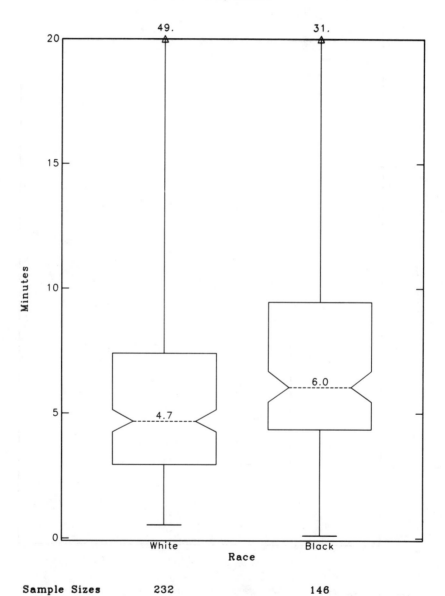

Box Plot Analysis of Local and Suburban Calling

Figure 5.B.7

Average Durations of Local Calls by Income

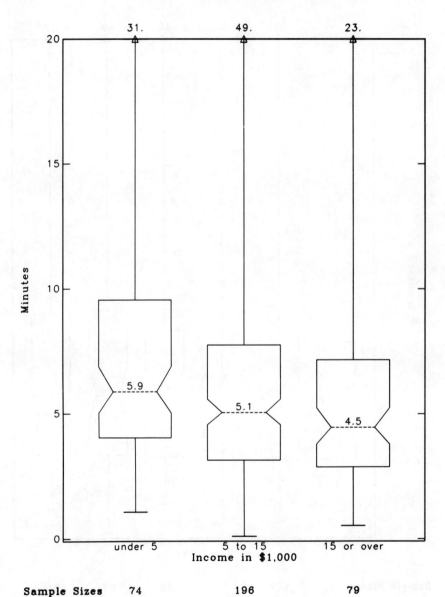

104 *The Effect of Demographics on Telephone Usage*

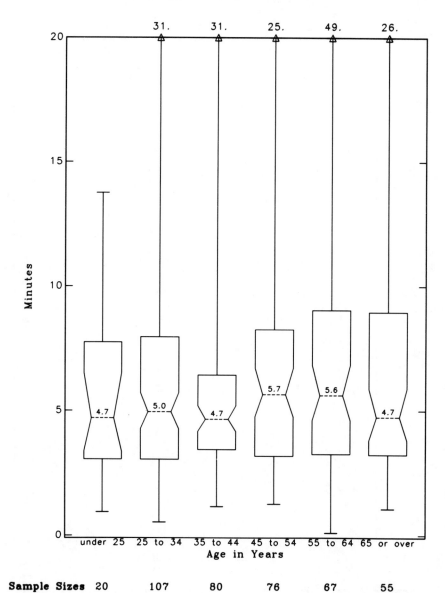

Figure 5.B.8

Average Durations of Local Calls
by Age of Head of Household

Figure 5.B.9

Total Local Conversation Time for Entire Sample

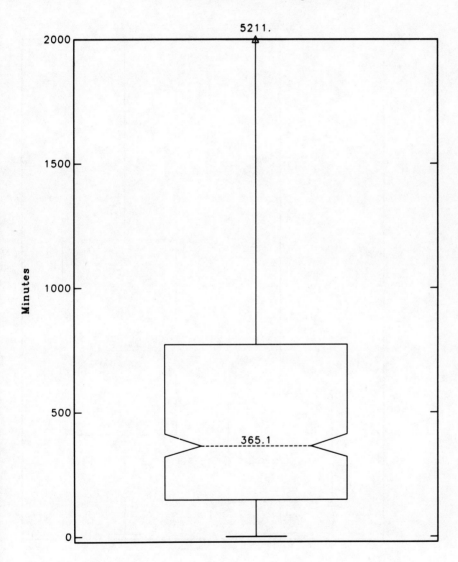

Sample Size 573

106 *The Effect of Demographics on Telephone Usage*

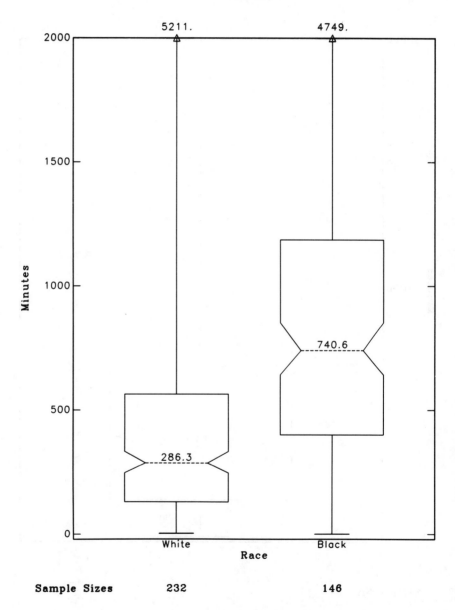

Figure 5.B.11

Total Local Conversation Time by Income

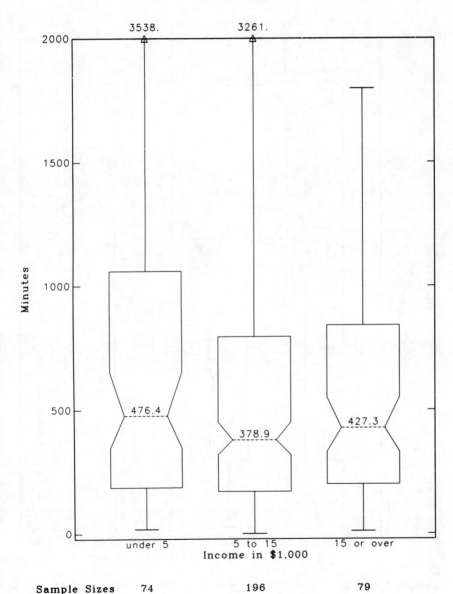

108 *The Effect of Demographics on Telephone Usage*

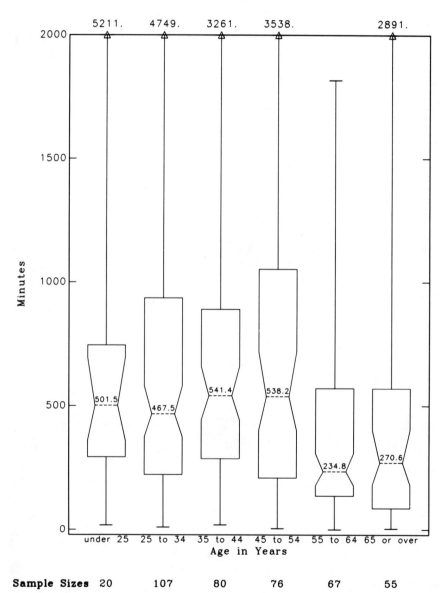

Figure 5.B.12

Total Local Conversation Time
by Age of Head of Household

Box Plot Analysis of Local and Suburban Calling

Figure 5.B.13

Observed vs. Customer Estimated Number of Local Calls

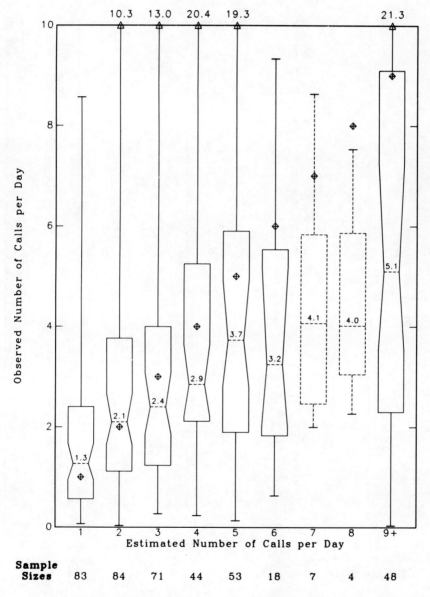

◆: Points where observed equals estimated number of calls

110 *The Effect of Demographics on Telephone Usage*

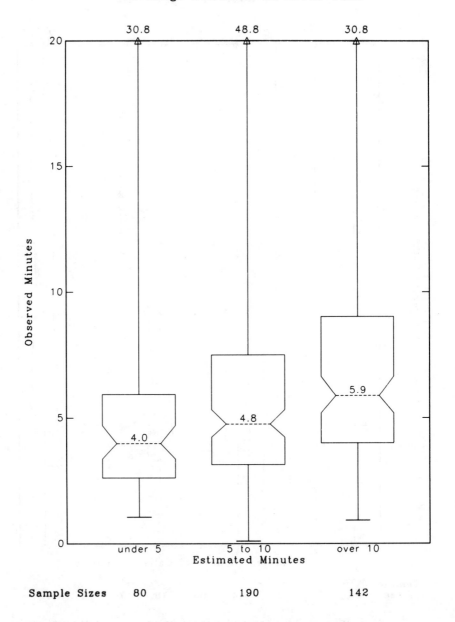

Figure 5.B.14

Observed vs. Customer Estimated
Average Duration of Local Calls

Figure 5.B.15

Number of Suburban Calls for Entire Sample

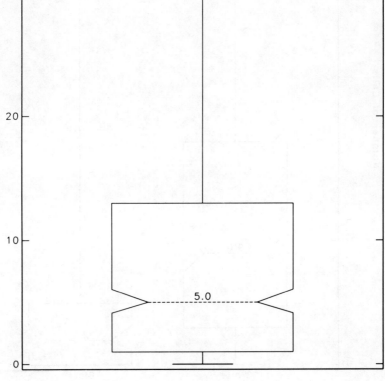

Sample Size 573

112 *The Effect of Demographics on Telephone Usage*

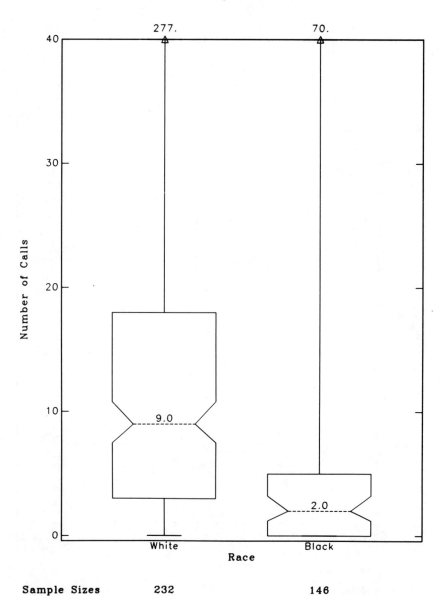

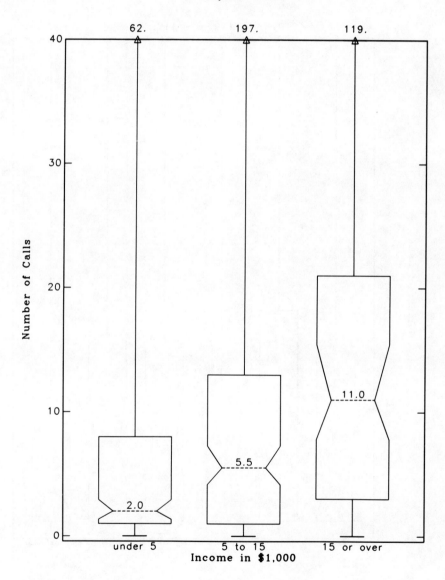

Figure 5.B.17

Number of Suburban Calls by Income

114 *The Effect of Demographics on Telephone Usage*

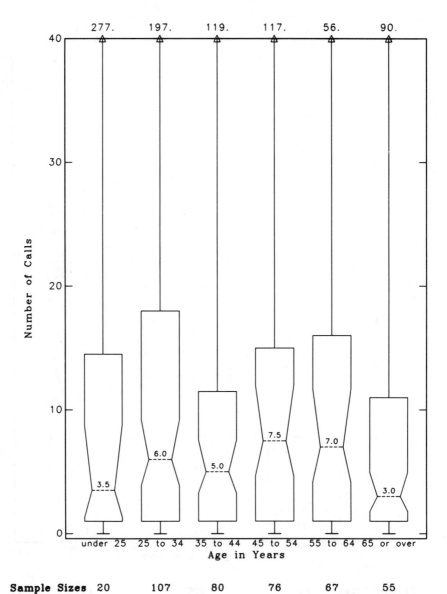

Figure 5.B.19

Average Durations of Suburban Calls
for Entire Sample

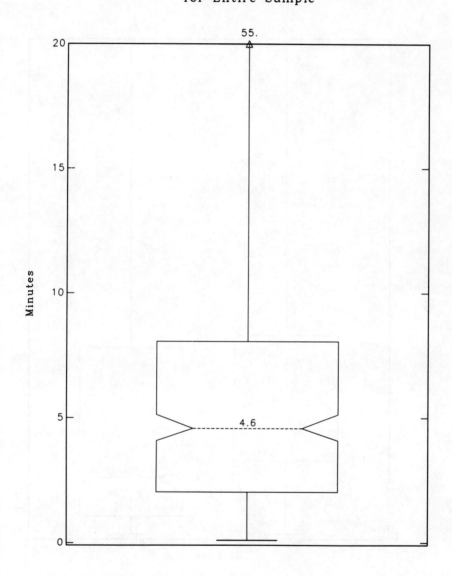

Sample Size 480

116 *The Effect of Demographics on Telephone Usage*

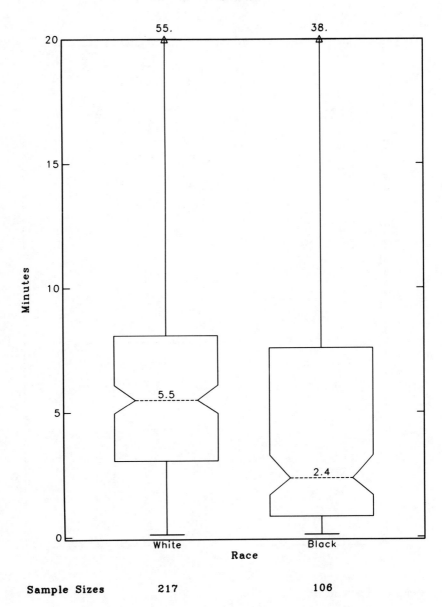

Figure 5.B.20

Average Durations of Suburban Calls by Race

Figure 5.B.21

Average Durations of Suburban Calls by Income

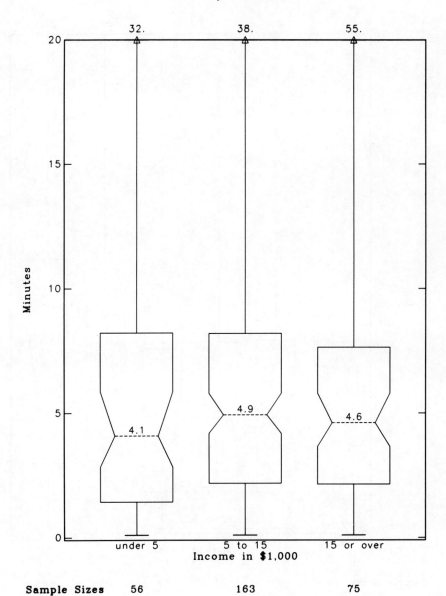

118 *The Effect of Demographics on Telephone Usage*

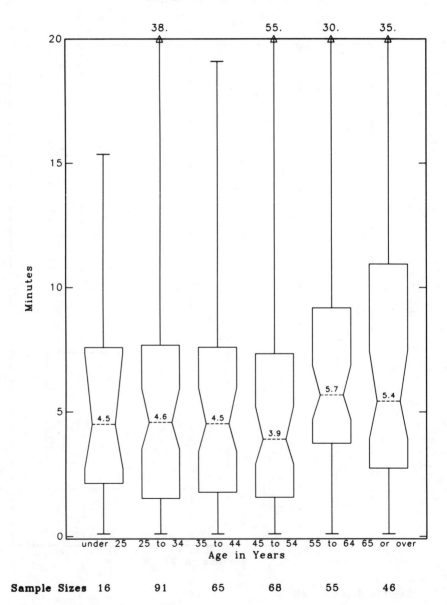

Figure 5.B.23

Total Suburban Conversation Time for Entire Sample

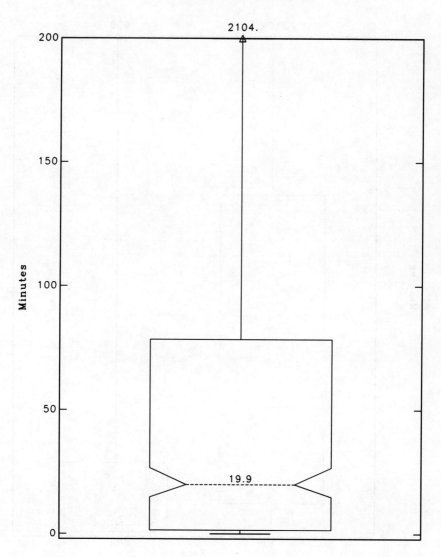

Sample Size 573

120 *The Effect of Demographics on Telephone Usage*

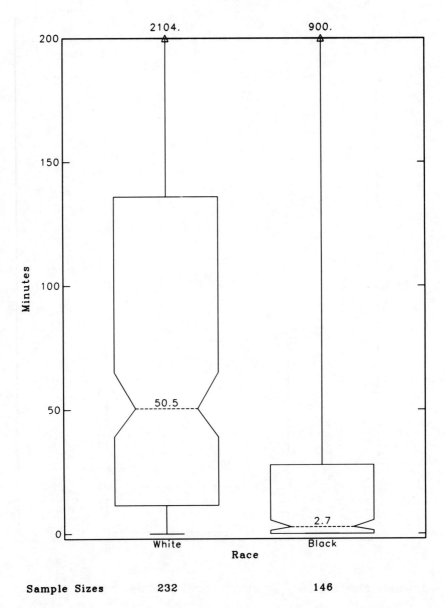

Figure 5.B.24

Total Suburban Conversation Time
by Race

Figure 5.B.25

Total Suburban Conversation Time by Income

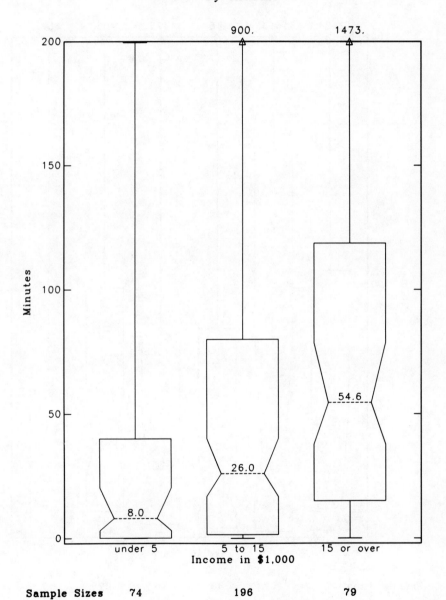

122 *The Effect of Demographics on Telephone Usage*

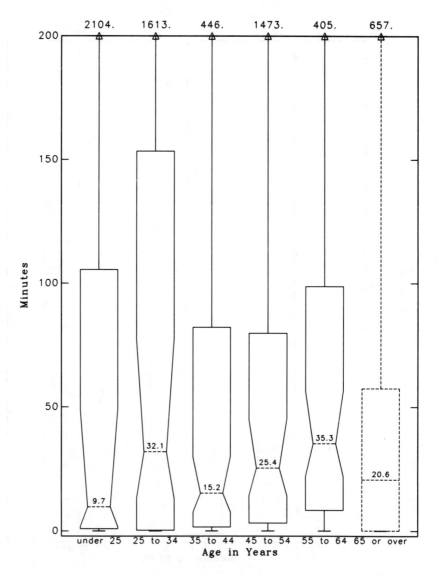

Figure 5.B.27

Suburban Message Units for Entire Sample

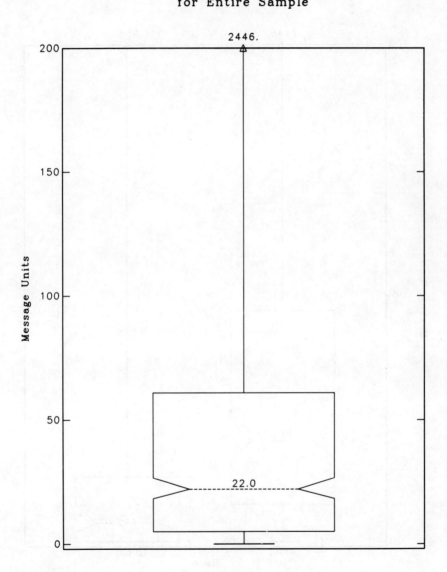

Sample Size 573

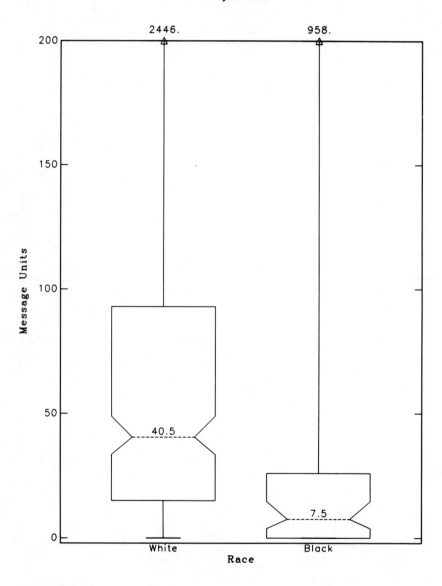

Figure 5.B.29

Suburban Message Units by Income

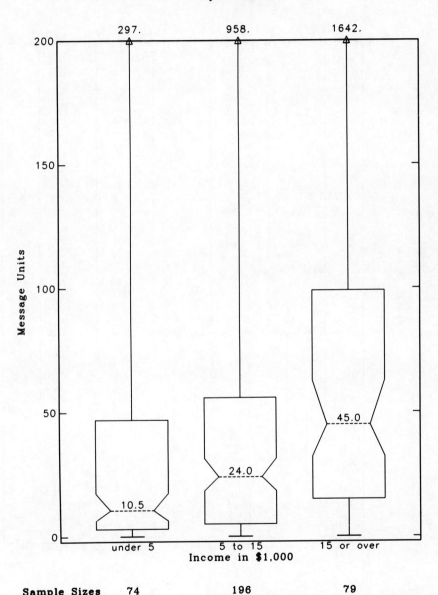

126 *The Effect of Demographics on Telephone Usage*

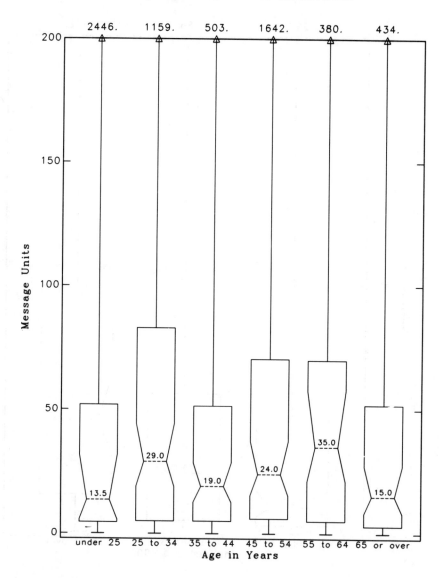

Figure 5.B.30

Suburban Message Units
by Age of Head of Household

Box Plot Analysis of Local and Suburban Calling 127

Figure 5.B.31

Total Local and Suburban Message Units for Entire Sample

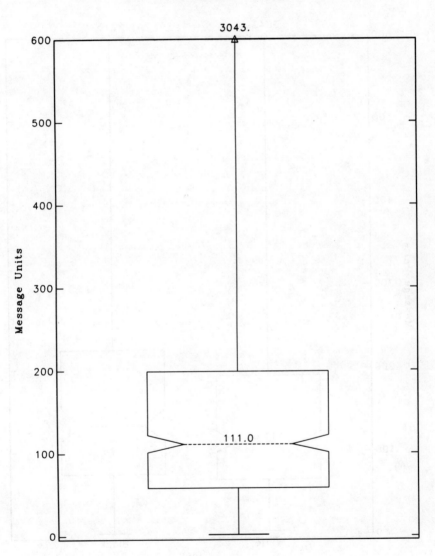

Sample Size 573

128 *The Effect of Demographics on Telephone Usage*

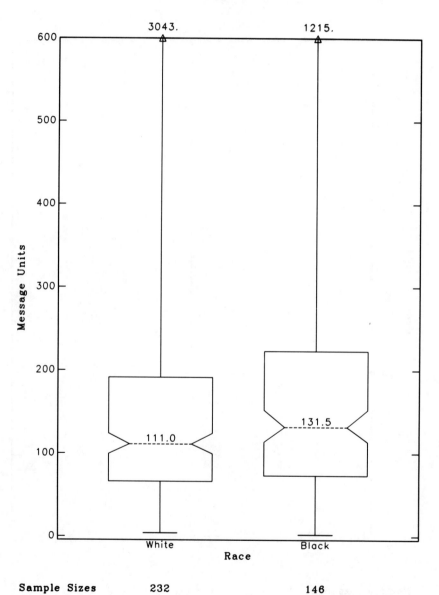

Figure 5.B.33

Total Local and Suburban Message Units by Income

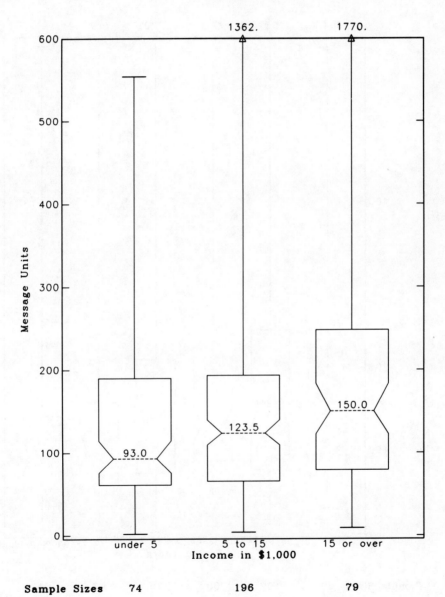

| Sample Sizes | 74 | 196 | 79 |

130 *The Effect of Demographics on Telephone Usage*

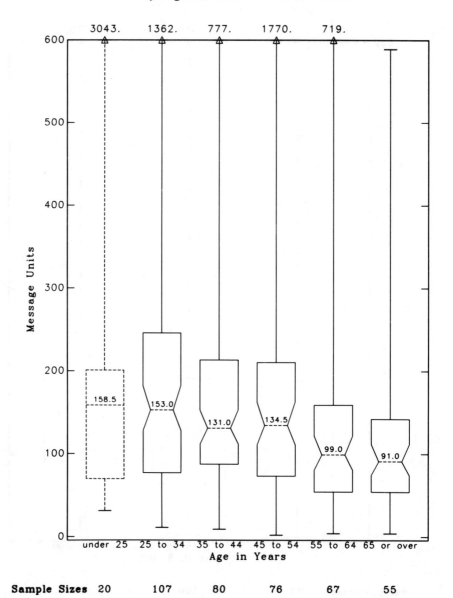

Box Plot Analysis of Local and Suburban Calling 131

Figure 5.B.35

Total Local and Suburban Message Units
by Message Unit Allowances
for Entire Sample

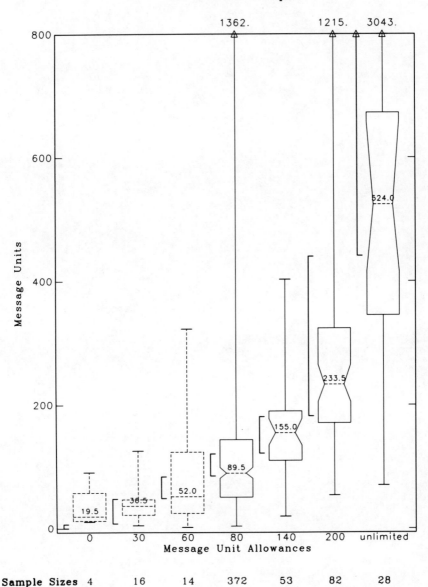

Note: Brackets represent the range of message units for which each allowance is the lowest-cost choice.

CHAPTER 6

Regression Analysis of Local and Suburban Usage

Paul S. Brandon [1]

6.1 Introduction

A variety of techniques are used in this book for analyzing the relationship between demographic characteristics and the usage of telephone services, including the comparison of medians via box plots, the comparison of means, the comparison of multinomial distributions, and Tobit regression analysis. The present chapter uses generalized least squares multiple regression analysis to analyze the non-toll calling of the sample of Chicago customers.

As Chapter 5 does, this chapter studies how household demographic characteristics are associated with the number of local calls, the average duration of local calls, total local conversation time, suburban

---

[1] Several persons have helped me in various ways during this investigation. I wish especially to acknowledge the ready and substantial assistance of Belinda B. Brandon in both technical and non-technical ways. I also wish to single out William E. Taylor for recognition for his strong technical help. Others who have helped are Elsa M. Ancmon, Richard A. Becker, Bruce C. N. Greenwald, Jerry A. Hausman, William J. Infosino, Roger W. Klein, Roger W. Koenker, Colin L. Mallows, Carl Pavarini, Mary H. Shugard, Richard G. Stafford, Stephanie Szabo, and Irma Terpenning. Only the author is responsible for any errors.

message units, and total local plus suburban message units.[2] The demographic groupings that are utilized are the age and sex of household members, race, income, the number of years of residence in Chicago, the employment status of the head and of the spouse, and the title or position of the head in his or her job.

Sets of variables representing the divisions of each grouping are used together in regressions. As a consequence, remember that the interpretation of the regression results is different from the interpretation when one is comparing medians or means across the divisions within a single grouping. For example, in the latter case, one addresses a question such as, "What is the difference in usage between the telephone customers with incomes of under $5,000 and of $5,000 to $8,999?" Instead the regressions in this chapter address a question such as, "What is the difference in usage between households that differ in having incomes of under $5,000 versus $5,000 to $8,999, *but that are in other respects the same?*" Thus, the regressions enable one to see whether a pattern in the means or medians that one observes across the divisions of some grouping may really be due to the correlation of those divisions with the divisions of another grouping that has a strong effect on usage. A multiple-grouping analysis is still subject to the bias of left-out variables, but there are fewer variables that are left out.

Of course, the explanatory and predictive power of a multiple-grouping analysis is greater than a single-grouping analysis. The emphasis in this chapter is on developing a model that is good for predicting the usage for customers outside the sample. (But it is not designed to predict beyond the period of estimation.)

In the second section of this chapter is a description of the data for the regressions and of the adjustments that are made to it. The third section explains the chosen econometric model and the procedure that is used to accommodate the correlation across months of the usage of each customer and to accommodate the pattern that high-usage groups have a greater variability in usage between customers and between months than do low-usage groups. The fourth section is devoted to displaying and discussing the regression results from the pooled, generalized least squares estimation. Some concluding remarks

---

[2] Recall the rate structure facing a telephone customer in Chicago: when he makes a local call, he incurs one message unit regardless of the duration or distance of the call. If he calls the suburbs, he may incur more than one message unit to initiate the call, and further message units related to the duration of the call; both the initial and duration-related message unit charges increase with distance.

and suggestions for further research comprise the final section. Readers without a technical orientation may be satisfied with skipping directly to the results in Section 6.4.

6.2 The Data

This section has four tasks. Firstly, it describes the sample that is used in the regressions. Secondly, it explains the adjustments that are made to the telephone usage data. Thirdly, it enumerates the variables that are used in the analysis. Fourthly, it describes some inferences that were made as to some missing questionnaire information.

The set of customers used for the regressions is drawn from the sample of Chicago customers for whom questionnaire data are available. Since Chapter 5 also studies local and suburban usage, the differences between the set of customers used in that chapter and the present one are detailed as follows: Firstly, the largest difference is that about 150 more customers enter the regressions than enter the analysis of Chapter 5. The reason is that the regressions use not only 1974 data but also summary calling data from 1972 and 1973, before many customers were disconnected. Secondly, unlike Chapter 5, sixteen customers are included for whom part of a month's calling data are missing in 1974 (their inclusion is possible because usage is expressed herein as rates per day). Thirdly, in the case of ten customers for whom detailed 1974 calling data are missing, total message unit usage is available on their bills. So these customers are included in the message unit regression, but not in Chapter 5. Fourthly, ten customers who are included in Chapter 5 are omitted from the regressions because of an excessive amount of missing questionnaire data; the criterion for omission is that there be data missing for two out of these three groupings: (1) income, (2) race, and (3) ages and sexes of household members. For these ten customers, the missing data could not be confidently inferred. Those three demographic characteristics provide much of the explanatory power of the regressions. It was judged not to be worth including these customers in the regressions, since additional variables would have had to have been created to represent the fact that data were not available.

For each customer in each month, the data on the number of local calls, total local conversation time, suburban message units, and total local and suburban message units are all divided by the number of equivalent days in the billing period for which the customer was observed. (According to the detailed 1974 data, the daily rate of usage is about 8 percent lower on weekends plus holidays than it is on weekdays, so a weekday is counted as one day and a weekend day or holiday is counted as 0.92 days.³) The other adjustment to the usage data is eliminating, from each customer's average recorded duration of local calls, the estimated bias due to rounding the recorded durations of calls up to integer minutes at tenths of minutes. It is done in a manner similar to that in Chapters 5, 7, and 8. The difference in the procedure between this chapter and those other chapters is that, because of a lack of detail in the data for 1972 and 1973, the operation can be performed here only for the aggregate average duration; and the estimate of the mean duration cannot be disaggregated by hour of the day as in those other chapters.

Data on telephone usage are drawn from all three available months. The summary data are used for October, 1972. For April, 1973, and April, 1974, the detailed data are used unless they are missing; if they are missing, then summary data are used. The summary data are used heavily for 1973, but only lightly for 1974. As mentioned before, message unit usage was taken from bills in some cases. For each measure of usage, the number of customers and the total number of observations are as follows:

Measure of Usage	Number of Customers	Number of Observations
Number of local calls	603	1526
Average duration of local calls	603	1525
Total local conversation time	603	1526
Suburban message units	602	1517
Total local and suburban message units	603	1530

Questionnaire data are all taken in early 1973. (See Chapter 3 for a description of the usage and questionnaire data.) The set of variables representing various demographic characteristics is listed as Exhibit 6.1 on the following page. A notation is included for whether each variable is an integer, a dummy variable, or a default variable. If it is a dummy

---

[3] A slight improvement over this procedure would have been to include terms in the regression such as the number of Sundays times the number of workers in the household. This suggestion, as well as several other useful comments and suggestions, came from William J. Infosino, Carl Pavarini, Mary H. Shugard, and Richard G. Stafford. The adjustment described in the text does not solve all the problems due to the variation in usage across days of the week. A further slight improvement in the model might result from one or more of the following procedures: (1) Use a different adjustment for each measure. (2) Treat each day of the week differently. (3) Estimate a different day-of-week pattern for various demographic groups.

Exhibit 6.1

List and Type of Variables

Demographic Characteristic	Form of Variable
Age and sex composition	
At least one child under 10 years	Dummy
Number of males 10-12 years	Integer
Number of males 13-15 years	Integer
Number of males 16-18 years	Integer
Number of males 19-24 years	Integer
Number of males 25-34 years	Integer
Number of males 35-54 years	Integer
Number of males 55-64 years	Integer
Number of males 65 years or older	Integer
Number of females 10-12 years	Integer
Number of females 13-15 years	Integer
Number of females 16-18 years	Integer
Number of females 19-24 years	Integer
Number of females 25-34 years	Integer
Number of females 35-54 years	Integer
Number of females 55-64 years	Integer
Number of females 65 years or older	Integer
Lives alone	
Alone	Dummy
Not alone	Default
Income	
Under $5,000	Default
$5,000 to $8,999	Dummy
$9,000 to $14,999	Dummy
$15,000 to $19,999	Dummy
$20,000 to $29,999	Dummy
$30,000 or more	Dummy
Income not available	Dummy

(Continued on next page)

Race

White	Default
Black	Dummy
Spanish speaking	Dummy
Other	Dummy

Years in Chicago

Less than one year	Dummy
1 year or more	Default

Employment status of head of household

Employed full time	Dummy
Employed part time	Dummy
Unemployed, looking for work	Dummy
Unemployed, not looking for work	Dummy
Retired	Default

Employment status of spouse (if married)

Employed full time	Dummy
Employed part time	Dummy
Unemployed, looking for work	Dummy
Unemployed, not looking for work, or retired	Default
Employment status not available	Dummy

Title or position of head (if employed)

Professional or technical	Dummy
Self employed, not specified	Dummy
Managerial	Dummy
Sales	Dummy
Craft	Dummy
Operative	Dummy
Transportation operative	Dummy
Laborer	Dummy

(Continued on next page)

Service or private household worker — Dummy
Occupation not available — Dummy
Clerical worker — Default

Interaction of race and income

Black × $5,000 to $8,999 — Dummy
Black × $9,000 to $14,999 — Dummy
Black × $15,000 to $19,999 — Dummy
Black × $20,000 or more — Dummy
Other races and incomes — Default

Interaction of number of people over 10 years and income

Number of people × $5,000 to $8,999 — Integer
Number of people × $9,000 to $14,999 — Integer
Number of people × $15,000 to $19,999 — Integer
Number of people × $20,000 to $29,999 — Integer
Number of people × $30,000 or more — Integer

Interaction of number of people over 10 years and race

Number of people × black — Integer

Control variables

Time index for 1972 — Integer
Time index for 1973 and 1974, except August, 1974 — Integer
Substitution of August for April, 1974 — Dummy
No substitution — Default
Data from 1972 — Dummy
Data from 1973 — Default
Data from 1974 — Dummy

variable, then it takes on a value of one for a household that has the listed characteristic and zero otherwise. If it is listed as a "default" variable, then it does not actually enter the regression as a variable but rather is the inferred characteristic of a household if none of the other dummy variables in its grouping equals one.

Some variables require special comment. (1) The variables for the number of persons of each age and sex were tried as logarithms, but produced a standard error of regression that was no smaller than that for the linear specification. It was found in preliminary runs that the dummy variable for "at least one child under 10 years" also performed better than separate or joint integer variables for the number of males and females under 10. (2) Similarly, in the case of "years in Chicago" it is advantageous to condense all of the several choices of a year or more into one variable. Also, please note that if in early 1973 a respondent had been in Chicago less than one year, then by April, 1974, he had been in Chicago for more than one year. So this variable takes on the value zero for all customers in the latter month. This is the only instance in which questionnaire information is varied across months. (3) Regarding the employment status of the spouse, it is presumed that "unemployed, not looking for work" is indistinguishable from "retired" from the point of view of generating calls. The results from preliminary regressions, in which those two categories were separate, supported this presumption. (4) The set of interactions between race and income is restricted to have five categories. The cell sizes would be too small to investigate a race-income interaction for Spanish-speaking or "other" households; and there are too few households that are black with an income of $30,000 or more to justify a separate variable for that group. (5) The time index for 1972 is the number of days later than August 31, 1972, that the billing period for a customer begins. In 1973 and 1974, the index is measured from March 31. These indices are intended to control for seasonality. The 1973 and 1974 indices are constrained to have the same coefficient because seasonality is assumed to remain constant from one year to the next. Note that August, 1974, is singled out to have its own dummy variable. The linear effect of the other time indices is chosen as an approximation. August is so much later than the other months for which data are available in 1974 that, if it were treated like the others, it might distort the coefficient on the 1973-1974 time index by stretching the ability of the linear specification to approximate the seasonal pattern. (6) Education of the head and of the spouse are not among the list of variables. They were tried in early runs, but there was no systematic pattern of effects across education levels.

Approximately 70 alterations are made in the received questionnaire data for the regressions. Virtually all of the changes are the elimination of cases in which an answer to a question in the questionnaire was not given by the respondent. Inferences were made of what the missing answer should have been, and the inference is used as data in the regressions. There is a variety of information in a questionnaire that can be used to infer with a fair degree of confidence what a missing answer should be. For example, a respondent might fill in information about himself in the table on the age and sex of household members, yet give no answer when asked the age and sex of the head; the first can be used to fill in the second. As another example, there were a variety of cases in which the head was married, yet no information was filled in on the employment status of the spouse. It was presumed that much of the time respondents did not consider it to be appropriate to check off "unemployed, not looking for work" or "retired" when the woman of the house was a housewife; so whenever there were small children in the house, and the head was employed, the spouse's employment status was changed to "unemployed, not looking for work" if no response had been given for it before; or if the spouse were 65 years of age or older, then her "not available" entry was changed to "retired" (but recall that no differentiation is made between "unemployed, not looking for work" and "retired" in the regressions). These changes to the employment status of the spouse comprise over half of the changes. There are a variety of other similar kinds of inferences.

6.3 The Econometric Model and Estimation Procedure

This section explains the econometric model that is chosen to describe the data. It also presents the procedure for the estimation of the model.

The model that relates the demographic and control variables to each of the measures of telephone usage is hypothesized to be linear. Let the subscripts i and t index the variables for customer i in month t, respectively, where i ranges from 1 to N, and t takes on one, two, or all three of the values 1, 2, and 3, depending upon which of the three months of usage data are available for the customer. As shorthand, the set of months for which data are available for a customer i is denoted τ_i, and the number of such t's in τ_i is T_i. The following notation is also used: y_{it} denotes any one of the telephone usage measures for customer i in month t; x_{jit} denotes the jth independent variable (*i.e.*, demographic or control variable) for that customer and month, where $j = 1,...,k$; β_j denotes the coefficient for the jth variable, with β_0 being the constant term; and w_{it} is an error term. Using this notation, the true model is assumed to be

$$y_{it} = \beta_0 + \beta_1 x_{1it} + \cdots + \beta_k x_{kit} + w_{it}, \qquad (6.1)$$

where $t \in \tau_i$, and $i = 1,...,N$. Letting x_{it} represent the $1 \times (k+1)$ vector of all variables for customer i in time t, and letting β represent the $(k+1) \times 1$ vector of coefficients, the model can be succinctly written as

$$y_{it} = x_{it}\beta + w_{it}. \qquad (6.2)$$

This linear specification has advantages and disadvantages. As stated in Section 6.1, the primary aim of this chapter is to develop a model that is good for predicting the usage for customers outside the sample. In this context, three of the advantages of the linear specification are as follows: Firstly, an investigation of alternative functional forms revealed that specifications in which the dependent variable was a power transform of usage measures were not able to produce predictions that were superior to those of the linear model: For a given month t and parameter $\lambda < 1$, the non-linear regression

$$\frac{y_{it}^{\lambda} - 1}{\lambda} = x_{it}\beta + w_{it} \qquad (6.3)$$

was run, yielding $\hat{\beta}$, the vector of estimated parameters, and $\hat{\sigma}^2$, the estimated variance of the error. From Eqn. (6.3),

$$y_{it} = (1 + \lambda x_{it}\beta + \lambda w_{it})^{1/\lambda}. \qquad (6.4)$$

Approximately unbiased predictions of the raw usage measures are obtained from[4]

$$\hat{y}_{it} = (1 + \lambda x_{it}\hat{\beta})^{1/\lambda} \qquad (6.5)$$
$$+ \hat{\sigma}^2 \frac{\lambda}{2}(\frac{1}{\lambda} - 1)(1 + \lambda x_{it}\hat{\beta})^{1/\lambda - 2}(1 - x_{it}(X'X)^{-1}x_{it}'),$$

where X is the matrix of all the observations of x_{it} stacked vertically. For the data at hand, the mean squared error of the predictions from Eqn. (6.5) is at least as large as that of the predictions from the linear model for each of the measures. A second advantage of the linear specification is that, while some interaction terms are included in the

---

[4] The expectation of the three-term Taylor expansion of the right-hand side of Eqn. (6.4) about $1 + \lambda x_{it}\beta$ equals the expectation of the three-term Taylor expansion of the first term on the right-hand side of Eqn. (6.5) plus the second term, where the true values are substituted for estimates.

list of variables in Exhibit 6.1, in the five regressions in Section 6.4 none of the interactions is statistically significant. These results provide little evidence against the hypothesis that usage and the independent variables are linearly related. Thirdly, conditional on the validity of the model, the linear specification gives unbiased predictions and their standard errors very easily.

There are at least two disadvantages of the linear specification. Firstly, the linear model has severe heteroscedasticity; for each usage measure, the variance of the errors increases as the estimated usage increases. This heteroscedasticity should be modeled and eliminated by generalized least squares. A power transform as in Eqn. (6.3), with λ at about a half or a third, would have removed most of the heteroscedasticity. Secondly, the errors of the linear model are not Normally distributed. There are a large number of small negative values and fewer, larger positive values. In such a case, the standard F- and t-tests, which are based on the Normal distribution for the errors, can lose validity. However, in this case it seems reasonable that the number of observations is so large that the estimated β's in repeated samples should be very close to Normally distributed.

Presumably, both the heteroscedasticity and the skewed errors are due to the fact that usage cannot be negative. Ideally one should like to model the probability distribution of the errors explicitly and estimate the parameters by maximum likelihood. However, in this instance where data from different months are pooled, such a procedure is complicated by the fact that the errors are strongly correlated across months for each customer. This situation is usually modeled by assuming that there are two error components as follows:

$$y_{it} = x_{it}\beta + u_i + v_{it}, \qquad (6.6)$$

where the u_i's are the errors across customers that remain constant over time, and the v_{it}'s are the errors across months for a given customer. In this model, we have

$\operatorname{var} u_i \equiv \sigma_{u_i}^2,$

$\operatorname{var} v_{it} \equiv \sigma_{v_{it}}^2,$

$\operatorname{cov} u_i v_{it} = 0$ for all i and t,

$\operatorname{cov} u_i u_j = 0$ for all $i \neq j$,

$\operatorname{cov} v_{it} v_{jt} = 0$ for all $i \neq j$ and all t.

Consequently,

$$cov(u_i+v_{it})(u_j+v_{jt}) = 0 \text{ for all } i \neq j \text{ and all } t,$$
$$cov(u_i+v_{is})(u_i+v_{it}) = \sigma_{u_i}^2 \text{ for all } s \neq t,$$

and

$$var(u_i+v_{it}) = \sigma_{u_i}^2 + \sigma_{v_{it}}^2. \tag{6.7}$$

The distribution of the u_i's is such that their mean is zero and, because $y_{it} \geq 0$, their density becomes zero at a value no lower than $-x_{it}\beta$. (Note the difficulty that the u_i's are supposed to be constant across months, yet their lower bound varies with t.) The distribution of the v_{it}'s also has a mean of zero, but a lower bound of $-(x_{it}\beta+u_i)$. The variances of the u_i's and v_{it}'s increase with estimated usage.

Returning to the issue of maximum likelihood, the author was unable to find probability distributions that not only modeled the two error components well but also could be dealt with analytically in combination. A possibly fruitful area for future research may be to explore more thoroughly the probability distributions of the error components, perhaps also resorting to numerical integration in order to perform the maximum likelihood estimation.

The author decided to make no distributional assumptions, but to use a two-stage heteroscedastic, variance-components, generalized least squares procedure. Before describing the procedure, the stochastic assumptions of this study should be made explicit. The customers in the sample are randomly drawn, and they certainly satisfy the diFinetti-Hausman exchangeability criterion,[5] so the u_i and the v_{it} are assumed to be random. On the other hand, the systematic monthly variation is assumed to be comprised of fixed effects; that is, the estimates are made conditional on the three time periods for which data are available. The reasons for this decision are that seasonality is strong and not estimable from the few available months of data; systematic changes occurred during the sample period in important variables outside the model such as the real price of message units; and there is virtually no variation across months in the values of the available demographic variables to enable one to estimate a systematic monthly random error component.

The general outline of the estimation procedure is as follows: a regression is run, yielding the estimated values of the dependent

---

[5] Jerry A. Hausman, "Specification Tests in Econometrics", *Econometrica*, v. 46, no. 6 (November, 1978), pp. 1251-72.

variable $\hat{y}_{it} \equiv x_{it}\hat{\beta}$ and of the residuals $\hat{w}_{it} \equiv y_{it} - \hat{y}_{it}$. The two error components are estimated, and their functional relationships with estimated usage are modeled. Then the estimated variance of each component for each customer is used to perform generalized least squares (GLS) in the second stage.[6]

Now consider the details of the procedure. There are several estimators of variance components in the literature.[7] Maddala and Mount used Monte Carlo methods to compare estimators.[8] They calculated the mean squared errors of coefficients estimated in a second stage by GLS using the variance components estimated by each estimator from the residuals of a first stage. The results were close for all the two-stage estimators and the maximum likelihood estimator.

---

[6] If the variance of the error terms is a function of $x_i\beta$, then generalized least squares is not asymptotically efficient as compared to the maximum likelihood estimator, assuming that the distribution of the error terms is known and that the functional relationship between the variances and $x_i\beta$ is known. Even repeated iterations of this procedure does not produce an estimator that is asymptotically efficient under those conditions. This result is demonstrated by Ralph Braid, "An Evaluation of Estimation Techniques for Logit Probability Models When Applied to Aggregate Data on Choice Frequency", unpublished paper, Massachusetts Institute of Technology (November, 1978), generalizing a finding by Takeshi Amemiya, "Regression Analysis When the Variance of the Dependent Variable is Proportional to the Square of Its Expectation", *Journal of the American Statistical Association*, v. 68, no. 344 (December, 1973), pp. 928-34.

[7] Takeshi Amemiya, "The Estimation of the Variances in a Variance-Components Model", *International Economic Review*, v. 12, no. 1 (February, 1971), pp. 1-13; Pietro Balestra and Marc Nerlove, "Pooling Cross Section and Time Series Data in the Estimation of a Dynamic Model: The Demand for Natural Gas", *Econometrica*, v. 34, no. 3 (July, 1966), pp. 585-612; Charles R. Henderson, "Estimation of Variance and Covariance Components", *Biometrics*, v. 9, no. 2 (June, 1953), pp. 226-52; Yair Mundlak, "On the Pooling of Time Series and Cross Section Data", *Econometrica*, v. 40, no. 1 (January, 1978), pp. 69-85; Marc Nerlove, "Further Evidence on the Estimation of Dynamic Economic Relations from a Time Series of Cross Sections", *Econometrica*, v. 39, no. 2 (March, 1971), pp. 359-82; C. Radhakrishna Rao, "Estimating Variance and Covariance Components in Linear Models", *Journal of the American Statistical Association*, v. 67, no. 337 (March, 1972), pp. 112-15; Shayle R. Searle, *Linear Models* (New York: John Wiley and Sons, 1971); Thomas D. Wallace and Ashiq Hussain, "The Use of Error Components in Combining Cross Section With Time Series Data", *Econometrica*, v. 37, no. 1 (January, 1969), pp. 55-72.

[8] G. S. Maddala and T. D. Mount, "A Comparative Study of Alternative Estimators for Variance Components Models Used in Econometric Applications", *Journal of the American Statistical Association*, v. 68, no. 342 (June, 1973), pp. 324-28.

The author's choice of estimator for the present problem is that of Wallace and Hussain. Their estimator is extended here in three ways. Firstly, as revealed in the equations below, the estimator is altered by the fact that different customers are observed for different numbers of months. Secondly, it was desired to improve the estimated dependent variable from the first stage in the hope that the efficiency of the second stage should be improved.[9] Thus GLS is used in the first stage rather than ordinary least squares (which is efficient only if the variance of the customer error component is zero). In this first stage, the two error components are assumed to have equal, non-zero variances that are constant across months and across customers.

Let $y_{i\cdot}$, $x_{i\cdot}$, and $v_{i\cdot}$ represent the mean across months of the dependent variable, of the vector of independent variables, and of the monthly errors, respectively, for customer i. Then, from Eqn. (6.6), we have

$$y_{i\cdot} = x_{i\cdot}\beta + u_i + v_{i\cdot}. \qquad (6.8)$$

From Eqns. (6.6) and (6.8), it follows that

$$y_{it} - \theta_i y_{i\cdot} = (x_{it} - \theta_i x_{i\cdot})\beta + u_i(1-\theta_i) + v_{it} - \theta_i v_{i\cdot}, \qquad (6.9)$$

where θ_i is some constant. If θ_i is defined as

$$\theta_i \equiv 1 - \left[\frac{\sigma_v^2}{\sigma_v^2 + T_i \sigma_u^2}\right]^{1/2}, \qquad (6.10)$$

where T_i is the number of months for which customer i is observed, then it can be shown that the errors $u_i(1-\theta_i) + v_{it} - \theta_i v_{i\cdot}$ are independent across customers and across time and have constant variance. Thus, the generalized least squares estimator for β is obtained by applying ordinary least squares to the transformed observations represented by Eqn. (6.9). Under the assumption of equality of the two variances for the purpose of the first stage, Eqn. (6.10) becomes

$$\theta_i = 1 - (1+T_i)^{-1/2}. \qquad (6.11)$$

---

[9] It has been shown that, when the two error components are assumed to be homoscedastic in the true model, more precise estimates of the variances of the error components in the first stage do not necessarily result in a lower mean squared error of the estimates of a coefficient in the second stage. It is not known whether the same is true in the present heteroscedastic case. See William E. Taylor, "Pooling Time Series and Cross Section Data: Exact Finite Sample Results", *Journal of Econometrics*, forthcoming.

Wallace and Hussain treat the estimated residuals from the first stage as if they are the true errors. Ignoring the degrees of freedom lost by estimating the coefficients in the first stage results in biased estimates of the variances, but the bias is small in the present circumstance where there are so many observations relative to the number of variables.

The third extension of Wallace and Hussain's procedure is the relaxation of the assumption of homoscedastic errors. It is assumed that the variances of both error components for each customer are functionally related to the estimated dependent variable for that customer, averaged across months. The average is taken as a matter of necessity in the case of the customer error component u_i: if u_i is to be constant over time for each customer, then the variance of the u_i's must also be constant. The average is taken as a matter of convenience in the case of the monthly error component v_{it}, but since there is little variation in the independent variables across months, this simplifying assumption makes little difference. The notation for the variance of the monthly error is chosen to be $\sigma_{v_i}^2$, dropping the t subscript, to denote the fact that the variance is assumed not to vary over time.

Following Wallace and Hussain, if Eqn. (6.7) is the correct model, then

$$v_{it} - v_{i\cdot} = y_{it} - y_{i\cdot} - (x_{it} - x_{i\cdot})\beta . \qquad (6.12)$$

This equation is simply a rearrangement of Eqn. (6.9), with $\theta_i = 1$. The variance of the left-hand side of Eqn. (6.12) is

$$\mathrm{var}\,(v_{it} - v_{i\cdot}) = \sigma_{v_i}^2 \left[\frac{T_i - 1}{T_i} \right], \qquad (6.13)$$

so $\dfrac{T_i}{T_i - 1}(v_{it} - v_{i\cdot})^2$ is treated as data to be used in estimating $\sigma_{v_i}^2$ and its functional relationship with $x_{i\cdot}\beta$. The errors of average durations are modeled slightly differently from the other measures, and will be discussed below. For the remaining measures, the chosen functional form is

$$\left[\frac{T_i}{T_i - 1}(v_{it} - v_{i\cdot})^2 \right]^{1/10} = \gamma_0 + \gamma_1 x_{i\cdot}\beta + z_{it} , \qquad (6.14)$$

where z_{it} is an error term. The dependent variable in Eqn. (6.14) is transformed by the one-tenth power in order to make the residuals close to symmetric, reducing the distorting impact of the few very large values of $(v_{it} - v_{i\cdot})^2$. Remarkably, once that transformation is applied, a linear specification as per Eqn. (6.14) appears correct. (An illustration is in the appendix.) Substituting estimated values for the v_{it}'s and for β in Eqn. (6.14), the parameters γ_0 and γ_1 are estimated, as well as the second, third, and fourth moments of z_{it} (σ_z^2, σ_{z3}, and σ_{z4}, respectively). From Eqn. (6.14),

$$\frac{T_i}{T_i-1}(v_{it}-v_{i.})^2 = (\gamma_0+\gamma_1 x_i.\beta+z_{it})^{10}. \tag{6.15}$$

Calculating the 5-term Taylor approximation to the right-hand side of Eqn. (6.15) and letting the ^ denote estimates, the chosen estimator of the variance of the monthly error for customer i is[10]

$$\hat{\sigma}_{v_i}^2 = (\hat{\gamma}_0+\hat{\gamma}_1 x_i.\hat{\beta})^{10} + 45\,\hat{\sigma}_z^2(\hat{\gamma}_0+\hat{\gamma}_1 x_i.\hat{\beta})^8 \tag{6.16}$$
$$+ 120\,\hat{\sigma}_{z3}(\hat{\gamma}_0+\hat{\gamma}_1 x_i.\hat{\beta})^7 + 210\,\hat{\sigma}_{z4}(\hat{\gamma}_0+\hat{\gamma}_1 x_i.\hat{\beta})^6.$$

As mentioned above, average durations are handled a little differently from the other measures. The precision of a customer's average duration as an estimate of his true expected duration should increase as his number of calls increases. It is assumed that, for a given customer, the variance of his monthly error is inversely proportional to his number of calls. Then the estimating equation for the heteroscedasticity of the monthly errors of average durations, which is parallel to Eqn. (6.14) for the other measures, is

$$\left[\frac{T_i}{T_i-1}(v_{it}-v_{i.})^2 c_{it}\right]^{1/10} = \gamma_0 + \gamma_1 x_i.\beta + z_{it}, \tag{6.17}$$

where c_{it} represents the number of calls. Then the predictions of the variance of the monthly errors of average durations are, parallel to Eqn. (6.16),

$$\hat{\sigma}_{v_i}^2 = \frac{1}{c_{i.}}\left[(\hat{\gamma}_0+\hat{\gamma}_1 x_i.\hat{\beta})^{10} + 45\,\hat{\sigma}_z^2(\hat{\gamma}_0+\hat{\gamma}_1 x_i.\hat{\beta})^8 \right. \tag{6.18}$$
$$\left. + 120\,\hat{\sigma}_{z3}(\hat{\gamma}_0+\hat{\gamma}_1 x_i.\hat{\beta})^7 + 210\,\hat{\sigma}_{z4}(\hat{\gamma}_0+\hat{\gamma}_1 x_i.\hat{\beta})^6\right].$$

Note that, for simplicity, the number of calls is averaged across months for each customer so that $\sigma_{v_i}^2$ can be constant across months.

---

[10] The Taylor expansion includes terms with the moments of $\hat{\beta}$ and $\hat{\gamma}$, but $\hat{\beta}$ and $\hat{\gamma}$ are estimated sufficiently precisely that for simplicity such terms are ignored.

Next, the variance of the customer error u_i is estimated as a function of $x_i.\beta$ by the following procedure. From Eqn. (6.8),

$$u_i + v_i. = y_i. - x_i.\beta . \qquad (6.19)$$

The variance of the left-hand side is

$$\text{var}(u_i+v_i.) = \sigma_{u_i}^2 + \frac{\sigma_{v_i}^2}{T_i}. \qquad (6.20)$$

Thus, each $(u_i+v_i.)^2 - \dfrac{\sigma_{v_i}^2}{T_i}$ is an observation for estimating $\sigma_{u_i}^2$. The specification that is chosen to approximate the relationship of the variance of the customer error with $x_i\beta$ is

$$\log\left[(u_i+v_i.)^2 - \frac{\hat{\sigma}_{v_i}^2}{T_i} + \alpha\right] = \delta_0 + \delta_1|x_i.\beta| \qquad (6.21)$$

$$+ \delta_2[|x_i.\beta|-x_i.\beta]/2 + z'_{it}$$

where z'_{it} is an error term. The logarithm is taken on the left-hand side to make the z'_{it} closer to symmetric than otherwise, and α is added before the logarithm is taken because many of the estimated values of $(u_i+v_i.)^2 - \dfrac{\hat{\sigma}_{v_i}^2}{T_i}$ are negative. (Such negative values do not indicate the violation of assumptions of the model. They are the expected result of random sampling error.) The absolute value of $x_i.\beta$ is taken in the second term of the right-hand side because, for some measures of usage, the first stage of the linear model (6.6) produces a few negative $x_i.\beta$'s and the left-hand side of Eqn. (6.21) increases as the $x_i.\beta$'s get more negative. The third term in Eqn. (6.21) provides for the fact that the slope of the relationship between the variance of the customer error and $x_i.\beta$ is much steeper for negative values of $x_i.\beta$ than for positive ones, at least in the case of suburban message units; for the other measures, δ_2 is set to zero. Estimated values of the data are substituted into the left-hand side of Eqn. (6.21), and the parameters δ_0, δ_1, and δ_2, as well as the variance $\sigma_{z'}^2$, are estimated. From Eqn. (6.21),

$$(u_i+v_i.)^2 - \frac{\hat{\sigma}_{v_i}^2}{T_i} = \exp(\delta_0+\delta_1|x_i.\beta|+\delta_2[|x_i.\beta|-x_i.\beta]/2+z'_i) - \alpha , \qquad (6.22)$$

so the predicting equation for the variance of the customer errors is

$$\hat{\sigma}_{u_i}^2 = \left[1 + \frac{\hat{\sigma}_{z'}^2}{2}\right] \exp(\hat{\delta}_0 + \hat{\delta}_1 |x_i.\hat{\beta}| + \hat{\delta}_2 [|x_i.\hat{\beta}| - x_i.\hat{\beta}]/2) - \alpha \ . \quad (6.23)$$

The value for α is chosen to be slightly greater than the absolute value of the most negative of the observed values of $(u_i + v_i.)^2 - \frac{\hat{\sigma}_{v_i}^2}{T_i}$. The factor $\left[1 + \frac{\hat{\sigma}_{z'}^2}{2}\right]$ comes from taking the expected value of a 3-term Taylor expansion of the right-hand side of Eqn. (6.22); it removes almost all of the bias in the predicted variances that exists by virtue of the variances' being a non-linear function of a stochastic variable.[11]

Once the variances of the two error components are estimated for each customer as above, generalized least squares estimation is accomplished by performing ordinary least squares on the transformed data in the following equation:

$$\frac{y_{it} - \hat{\theta}_i y_i.}{\hat{\sigma}_{v_i}} = \frac{[x_{it} - \hat{\theta}_i x_i.]\beta}{\hat{\sigma}_{v_i}} + w'_{it} \ , \quad (6.24)$$

where

$$\hat{\theta}_i \equiv 1 - \left(\frac{\hat{\sigma}_{v_i}^2}{\hat{\sigma}_{v_i}^2 + T_i \hat{\sigma}_{u_i}^2}\right)^{1/2} . \quad (6.25)$$

The w'_{it} should have constant variance and should be independent across customers and across months. This GLS procedure produces an estimator for the coefficients that is consistent. The F- and t-tests are also asymptotically valid, and the sample is probably large enough that the asymptotic approximation is good.

A step-by-step illustration of the procedure is contained in Appendix 6.A. The number of local calls is used as the dependent variable for the illustration.

One final comment should be made regarding the procedure: since the author had a prior belief that usage of a household is influenced by its age and sex composition, income, and race, the sets of variables representing these groupings are forced into the regression. For both the first and second stages, the other variables are added, one at a time, in decreasing order of reduction in the sum of squared errors; additions

---

[11] Since the $\hat{\beta}$'s and $\hat{\delta}$'s are estimated fairly precisely, terms involving their second and higher moments are neglected.

are stopped when the mean square error is no longer improved (*i.e.*, when the next variable's estimated coefficient would fall short of its estimated standard error if the variable were added to the regression).

6.4 The Results

This section presents the results of the generalized least squares estimation of the linear specification with two heteroscedastic error components, following the procedure described in Section 6.3. Results are reported in Exhibit 6.2 for the number of local calls per equivalent day, the average duration per local call, total local conversation time per equivalent day, suburban message units per equivalent day, and total local and suburban message units per equivalent day. Where a particular variable does not enter the regression, a dash is entered in the table. Coefficients that are significant at the 5 percent level are shown in bold type, and the z-statistic for a Normal test is shown below each coefficient. In the case of the variables for income, a one-tailed Normal test is used for all usage measures except durations, since economic theory predicts that consumption of a priced normal good should rise with an increase in income, and since this tendency has been substantiated for a variety of other telephone services. But a two-tailed test is used for these variables for durations of local calls, since no price is attached to this dimension of usage in Chicago. For all other demographic variables the test is two-tailed.

Detailed results are discussed below, but it may be useful to make some general observations first. It is clear that demographic characteristics have an appreciable effect on telephone usage, as indicated by the size of the coefficients relative to the unweighted means of the usage measures near the top of the table. And there are a good number of significant coefficients. The values of the F-statistics in Exhibit 6.2 imply that each of the regressions is significant at better than the one percent level. (The critical values are all about 1.6.) If one concentrates on the demographic variables alone, as distinguished from the control variables, one finds that for each regression one can also reject the hypothesis that none of the demographic variables has any effect on usage at better than the one percent level. (The F-statistics are a little smaller than those in the Exhibit, but still exceed the critical values.) But it also appears that the available data on demographics account for only a small portion of the variation between individual households; the R^2's for the regressions range from 0.05 to 0.20.

Exhibit 6.2

Generalized Least Squares Regression Results

	Local			Suburban	Local and Suburban
Statistics	Number of Calls	Average Duration	Total Conv. Time	Message Units	Total Message Units
R^2	0.195	0.054	0.162	0.064	0.103
Regression degrees of freedom	38	35	38	36	37
Residual degrees of freedom	1487	1489	1487	1480	1492
F	9.51	2.34	6.52	2.70	4.27
Mean	3.48	6.17	20.26	1.92	5.42
Variables	Coefficients (with absolute values of z-statistics)				
At least one child under 10 years	**0.78**	-0.35	**3.23**	-0.06	0.65
	(3.34)	(0.86)	(2.03)	(0.25)	(1.58)
Number of males 10-12 years	-0.10	0.32	1.08	-0.41	-0.11
	(0.32)	(0.60)	(0.51)	(1.17)	(0.24)
Number of males 13-15 years	**0.77**	0.01	**4.40**	0.43	**1.52**
	(2.46)	(0.01)	(1.96)	(1.07)	(2.27)
Number of males 16-18 years	-0.18	-0.29	-0.94	0.35	-0.16
	(0.61)	(0.58)	(0.47)	(1.10)	(0.27)
Number of males 19-24 years	**0.72**	-0.09	**3.79**	0.01	1.41
	(2.69)	(0.20)	(2.14)	(0.05)	(0.63)
Number of males 25-34 years	0.25	-0.48	0.86	0.37	0.68
	(1.10)	(1.12)	(0.54)	(1.50)	(1.69)
Number of males 35-54 years	-0.09	-0.65	-1.52	0.14	0.37
	(0.36)	(1.42)	(0.91)	(0.53)	(0.84)
Number of males 55-64 years	-0.56	-0.28	-3.57	0.18	-0.23
	(1.87)	(0.45)	(1.87)	(0.53)	(0.44)
Number of males 65 years or older	0.44	-0.92	0.86	0.24	0.82
	(1.32)	(1.33)	(0.37)	(0.66)	(1.39)
Number of females 10-12 years	**0.97**	0.14	**4.49**	**-0.77**	0.41
	(2.92)	(0.28)	(2.00)	(2.73)	(0.79)
Number of females 13-15 years	**1.28**	-0.26	**5.29**	0.02	**1.46**
	(3.94)	(0.57)	(2.47)	(0.07)	(2.62)
Number of females 16-18 years	0.29	0.96	**5.39**	0.03	0.42
	(0.99)	(1.82)	(2.49)	(0.08)	(0.74)
Number of females 19-24 years	0.45	-0.23	1.83	0.16	0.82
	(1.73)	(0.50)	(1.04)	(0.65)	(1.77)
Number of females 25-34 years	0.42	0.65	**3.55**	**0.58**	**1.07**
	(1.62)	(1.29)	(1.96)	(2.17)	(2.28)
Number of females 35-54 years	0.31	0.58	**4.18**	0.40	0.55
	(1.19)	(1.18)	(2.36)	(1.62)	(1.22)
Number of females 55-64 years	-0.07	**1.47**	3.10	-0.04	-0.26
	(0.25)	(2.39)	(1.57)	(0.13)	(0.53)
Number of females 65 years or older	-0.16	**1.55**	-0.41	-0.03	-0.47
	(0.54)	(2.46)	(0.20)	(0.10)	(0.93)

(Continued on next page)

Lives alone	-0.47	0.81	-3.09	—	-0.82
	(1.73)	(1.47)	(1.68)		(1.73)
$5,000 to $8,999	-0.39	-0.85	-3.29	0.74	0.34
	(0.80)	(1.00)	(1.00)	(1.46)	(0.68)
$9,000 to $14,999	-0.38	0.54	0.36	0.25	-0.02
	(0.89)	(1.22)	(0.16)	(0.85)	(0.04)
$15,000 to $19,999	0.52	-0.45	3.91	0.47	-0.31
	(1.39)	(0.37)	(1.45)	(1.04)	(0.23)
$20,000 to $29,999	0.19	0.47	0.47	-1.09	0.06
	(0.40)	(0.56)	(0.14)	(0.84)	(0.06)
$30,000 or more	**1.93**	-1.61	3.43	**1.84**	**3.57**
	(3.39)	(1.72)	(0.92)	(2.27)	(3.03)
Income not available	**0.69**	0.56	2.38	—	0.99
	(2.19)	(1.68)	(1.07)		(1.86)
Black	**1.56**	**2.11**	**16.11**	**-1.26**	0.42
	(7.80)	(3.14)	(6.48)	(5.47)	(1.23)
Spanish speaking	-0.54	-1.07	**-4.55**	**-1.05**	**-1.42**
	(1.55)	(1.59)	(2.03)	(2.84)	(2.49)
Other races (but not white)	0.73	1.53	**8.29**	-0.71	-0.05
	(1.38)	(1.45)	(2.15)	(1.21)	(0.05)
Less than one year in Chicago	—	—	-4.97	—	—
			(1.50)		
Head employed full time	—	—	**-5.46**	—	—
			(2.11)		
Head unemployed, looking for work	0.67	—	—	—	—
	(1.16)				
Head unemployed, not looking for work	**1.09**	—	—	0.59	**1.71**
	(2.69)			(1.68)	(2.47)
Spouse employed full time	-0.45	—	**-3.69**	**-0.57**	**-1.24**
	(1.92)		(2.40)	(2.29)	(3.03)
Spouse unemployed, looking for work	—	—	—	—	-1.52
					(1.87)
Spouse employment status not available	—	—	—	-0.87	-1.45
				(1.29)	(1.17)
Professional or technical	—	0.63	2.96	—	—
		(1.18)	(1.63)		
Self employed	**2.33**	**-3.24**	—	—	4.24
	(1.99)	(2.00)			(1.65)
Managerial	—	-0.95	—	—	—
		(1.52)			
Sales	0.83	—	3.95	**1.73**	**2.74**
	(1.83)		(1.25)	(2.54)	(2.81)
Craft	0.30	—	1.81	—	—
	(1.10)		(1.01)		
Service or private household worker	0.40	—	—	—	—
	(1.28)				
Occupation not available	—	—	-3.02	—	—
			(1.26)		
Black × $5,000 to $8,999	—	—	—	0.78	—
				(1.69)	
Number of people × $5,000 to $8,999	0.24	0.30	1.64	-0.29	—
	(1.47)	(1.10)	(1.42)	(1.87)	

(Continued on next page)

Number of people × $9,000 to $14,999	0.18 (1.24)	—	—	—	—
Number of people × $15,000 to $19,999	—	0.46 (1.19)	—	—	0.71 (1.42)
Number of people × $20,000 to $29,999	—	—	—	0.59 (1.50)	—
Number of people × black	—	-0.40 (1.79)	-1.33 (1.45)	—	—
Time index for 1972 ÷ 100	**-0.53** (3.52)	—	—	—	—
Substitution of August for April, 1974	—	**1.18** (2.63)	—	**-0.78** (1.99)	**-1.65** (2.41)
Data from 1972	—	—	**-1.34** (2.81)	**-0.21** (2.03)	**-0.49** (3.07)
Data from 1974	0.12 (1.60)	-0.12 (1.14)	0.56 (1.13)	**0.22** (2.08)	**0.37** (2.26)
Constant term	**1.49** (3.48)	**5.16** (6.97)	**12.69** (3.56)	**1.35** (3.84)	**3.26** (4.64)

Of course, it may still be possible to predict aggregates such as neighborhoods or cities with good accuracy. The estimated variance of the error of prediction of the mean usage of some finite population is

$$var = x'\hat{V}x + \frac{1}{N^2} \sum_{i=1}^{N} \hat{\sigma}_i^2, \qquad (6.26)$$

where x is the column vector of mean demographic characteristics of the population, $\hat{V}$ is the estimated covariance matrix of the estimated coefficients, N is the number of households in the population, and $\hat{\sigma}_i^2$ is the estimated variance of the error term for household i. If the mean demographic characteristics remain constant as the size of the population becomes larger, then the first term on the right-hand side is constant as the second term becomes smaller. In particular, $x'\hat{V}x$ is the estimated variance of prediction for a very large population. For illustration, the standard errors of predicted means from the regression are shown in Exhibit 6.3 for a variety of sizes of populations. Such standard errors are shown for each of the five usage measures. In each case, usage is predicted for April, 1973, and the population is assumed to have the same demographic composition as the sample from which the regressions are estimated (in general, a prediction for a population with a different demographic composition would have larger variance). It is apparent that the prediction error is large for individual households, but that it is small for large populations.

Another general observation is that for the present data individual coefficients are moderately correlated with one another. It probably is not worth presenting the large covariance/correlation matrix of the

Exhibit 6.3

Standard Errors of Predicted Means

	Prediction of Mean	Number of Customers				
		1	10	100	1000	∞
Measure		Standard Error				
Number of Local Calls per Day	3.44	1.88	0.60	0.21	0.11	0.09
Average Duration per Local Call	6.12	6.09	1.93	0.63	0.25	0.16
Total Local Conversation Time per Day	19.78	15.34	4.89	1.66	0.80	0.64
Suburban Message Units per Day	1.59	1.64	0.53	0.20	0.12	0.11
Total Message Units per Day	5.24	2.85	0.92	0.34	0.20	0.18

coefficients for each regression but, in summary, the frequency distribution of the 741 coefficient correlations (excluding the constant term) for the number of calls is as follows:

Correlation	Frequency	Percentage
-0.8 to -0.7	2	0.3
-0.3 to -0.2	7	0.9
-0.2 to -0.1	45	6.1
-0.1 to 0.0	278	37.5
0.0 to 0.1	289	39.0
0.1 to 0.2	60	8.1
0.2 to 0.3	36	4.9
0.3 to 0.4	17	2.3
0.4 to 0.5	6	0.8
0.5 to 0.6	1	0.1

In the case of a variable whose coefficient has a high correlation with others, the estimated values of those other coefficients will be unstable with respect to the choice of whether or not the variable is put into the regression. Still, predictions of the regression will not display this instability.

Now consider the detailed results. Besides qualitative findings, only coefficients for household characteristics are mentioned that are statistically significant at the 5 percent level or better. Control variables are not discussed.

6.4.1 Number of Local Calls

The results for the regression in the left-most column of figures in Exhibit 6.2 are as follows: (1) The number of local calls that a household generates tends to increase with the number of people in the household (holding other things constant), but it matters what the ages and sexes of the household members are. The number of calls is especially affected by the presence of girls in the household who are 13 to 15 years old; the calling rate is estimated to be 1.3 per day higher for each such person. In order of decreasing importance, other age/sex groups with significant coefficients are girls 10 to 12 years, a child under 10, boys 13 to 15, and men 19 to 24. (2) The calling rate is 1.9 per day higher for the group with an income of $30,000 or more than for the group with an income of under $5,000. Imagine a set of households that are average in all demographic respects but income, where "average" is according to the demographic composition of the sample. The portion of this set with an income of over $30,000 is predicted to have a calling rate of 4.3 calls per day, almost 80 percent more than the predicted level of 2.4 calls for the under-$5,000 group. (3) The largest z-statistic is estimated for the difference in usage between black and white customers; the former make 1.6 more calls per day than the latter, when all other variables are held constant. (4) The number of calls is estimated to be higher if the head is unemployed and not looking for work than if he or she is employed or retired. (5) The only title or position of the head that comes out strongly is "self employed". Self employment is estimated to have a stronger effect on the number of calls — 2.3 calls per day — than does any other variable.

6.4.2 Average Duration of Local Calls

The results for durations are weaker than for the other measures. (1) Durations of local calls are significantly increased (by about a minute and a half) by the presence of women of ages 55 or over. (2) Black customers have 2.1 minutes longer durations than comparable white customers. (3) The only title or position of the head that comes out strongly is "self-employed". Self employment is estimated to have a stronger effect on durations than any other variable does — 3.2 minutes per call less.

6.4.3 Total Local Conversation Time

(1) Total local conversation time is estimated to be significantly increased by the presence of at least one child under 10 years old, boys 13 to 15, men 19 to 24, girls 10 to 18, and women 25 to 54. (2) The variable with the largest effect is race: the coefficient for black households indicates 16 more minutes per day than white households, representing a difference of almost 70 percent for households that are average in all respects except race; Spanish-speaking households spend less time on the telephone making outgoing local calls than white households; and "other" races spend more time. (3) Time spent on the telephone is also less if the head or spouse is employed full time.

6.4.4 Suburban Message Units

(1) For suburban message units, one of the age/sex variables — girls 10 to 12 years old — has a surprising negative sign; this instance is the only significant anomolous result for any measure of usage. More reasonable is the higher usage for women between 25 and 34 years old. (2) The highest-income group makes 1.8 more suburban message units than does the lowest-income group. For households that are otherwise average, the usage of the highest-income group is more than twice that of the lowest-income group. (3) Households that are either black or Spanish-speaking have significantly lower suburban usage than white ones; the difference is almost a factor of two for otherwise-average households. (4) A full-time job held by the spouse lowers suburban calling. (5) If the head of the household is in sales, then suburban usage is higher relative to households with heads of other positions.

6.4.5 Total Message Units

(1) As for total local and suburban message units, usage is significantly increased by the presence of boys or girls who are 13 to 15 years old and of women who are between the ages of 25 and 34. (2) Message unit usage is significantly higher for the highest-income group than for the lowest-income group. (3) Spanish-speaking customers have lower usage than white customers. (4) The fact that the head is unemployed and not looking for work appears to increase calling. A full-time job held by the spouse reduces calling. (5) Finally, a sales position increases message unit usage.

6.5 Conclusion and Suggestions for Future Research

In this as well as in other chapters, the demographic characteristics of households have been demonstrated to have a significant influence on the use of telephone services. But it is also clear that there must be many other factors that affect usage. The regression results for each measure are qualitatively similar to the results of the box plot analysis of Chapter 5, even though Chapter 5 compares medians (not means) for only one demographic characteristic at a time. The primary difference between the patterns in this chapter and in the previous one is that here the income effect is weaker and less smooth.

As mentioned above, these regressions are designed for prediction outside the sample. A useful test of the model would be to compile data on the demographic composition of telephone subscribers in the areas of Chicago that were not sampled for this study to see whether the regressions predict their usage well.

As mentioned in Section 6.3 above, it would be desirable if the probability distribution of the error terms could be modeled well, so that the maximum likelihood estimation procedure could be applied. Under that circumstance, the model for suburban message units could be improved by using something like Tobit analysis, which recognizes the non-negligible fraction of the sample that makes no suburban calls.

There is another technical issue that might be fruitfully investigated — the question of whether bias is imparted to the estimates by data's being missing.

Other suggestions for future research that are of broader scope are made in Chapter 11.

Appendix 6.A

Illustration of Generalized Least Squares Procedure

In this appendix the detailed two-stage generalized least squares procedure that is described in Section 6.3 is illustrated for one of the usage measures — the number of local calls. In addition, the results of that estimator are compared with the results of three other estimators, and the hypothesis that coefficients are constant across months is tested.

The full pooled heteroscedastic error components procedure is here denoted "PHEC". From the first stage, as per Eqns. (6.9) and (6.11) in the main body of the chapter, come the $x_i.\beta$'s and the estimated values of $v_{it} - v_i.$. The plot of each estimated value of $(v_{it}-v_i.)\left[\dfrac{T_i}{T_i-1}\right]^{1/2}$ versus $x_i.\hat\beta$ is shown as Figure 6.A.1 on the following page. A pronounced increase in the spread of the errors with larger values of $x_i.\beta$ is visible. The regression relating the variance of the monthly errors to $x_i.\beta$, specified by Eqn. (6.14), is

$$\left[\dfrac{T_i}{T_{i-1}}(v_{it}-v_i.)^2\right]^{1/10} = \underset{(68.91)}{0.775} + \underset{(10.63)}{0.032}\; x_i.\hat\beta + \hat z_{it}, \quad (6.A.1)$$

with $F(1, 1489) = 112.93$, $R^2 = 0.070$, $\hat\sigma_z^2 = 0.0395$, $\hat\sigma_{z3} = -0.0017$, and $\hat\sigma_{z4} = 0.0051$ (t-values for the coefficients are in parentheses). The very large t-value for $x_i.\beta$ implies that the hypothesis of homoscedasticity for the monthly error component can be rejected at better than the 0.1 percent level. The plot of the dependent variable versus the independent variable together with the regression line for Eqn. (6.A.1) is shown in Figure 6.A.2. Note the adequacy of the straight line, the symmetry of the scatter of points around the regression line, and the lack of outliers.

Next, the estimated squared customer residuals are calculated. The minimum estimated $\sigma_{u_i}^2$ is -2.76, so 2.8 is added to each estimate to make them all positive. The estimate of Eqn. (6.21) is then

$$\log\left[(u_i+v_i.)^2 - \dfrac{\hat\sigma_{v_i}^2}{T_i} + \alpha\right] = \underset{(14.87)}{1.165} + \underset{(4.59)}{0.094}\; |x_i.\hat\beta| + \hat z_{it}', (6.A.2)$$

where $\hat\sigma_{v_i}^2$ is estimated from Eqn. (6.18). For this regression, $F(1, 586) = 21.08$, $R^2 = 0.037$, and $\hat\sigma_{z'}^2 = 0.704$. The large t-value on

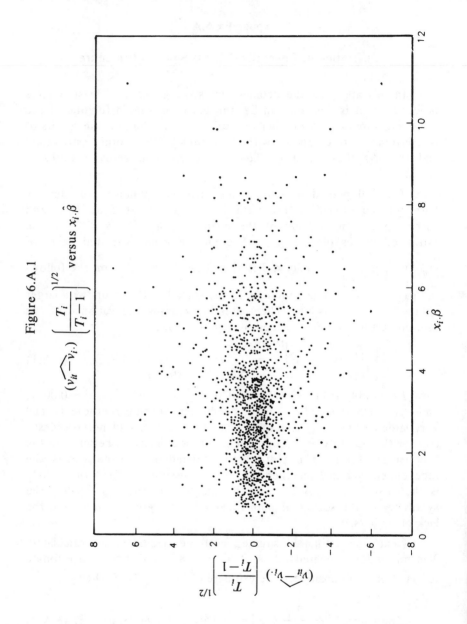

Figure 6.A.1 $(\widehat{v_{it} - v_{i\cdot}}) \left[\dfrac{T_i}{T_i - 1} \right]^{1/2}$ versus $x_i \hat{\beta}$

Regression Analysis of Local and Suburban Usage 161

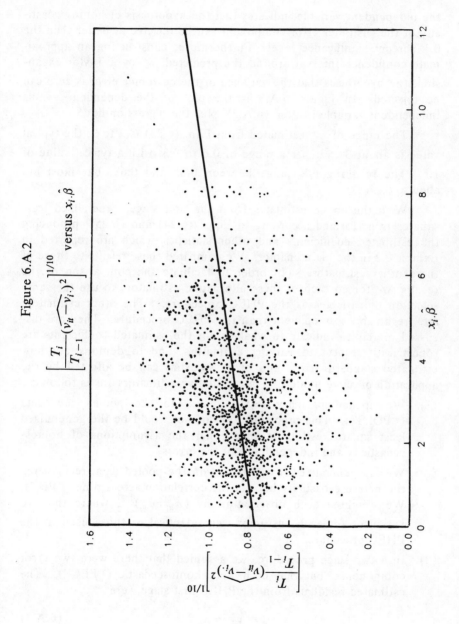

Figure 6.A.2 $\left[\dfrac{T_i}{T_{i-1}}\widehat{(v_{it}-v_{i\cdot})^2}\right]^{1/10}$ versus $x_i.\hat{\beta}$

the independent variable indicates that the hypothesis of homoscedasticity of the customer error component can be rejected at better than the 0.1 percent significance level. Furthermore, constructing an approximate confidence interval around the predicted $\sigma_{u_i}^2$ by a Taylor expansion, the hypothesis that the variance of the customer error is zero can be rejected. In Figure 6.A.3 is the plot of the dependent versus independent variable in Eqn. (6.A.2), plus the regression line.

The range of $\hat{\sigma}_{u_i}^2$ estimated from Eqn. (6.23) is 2 to 9; the typical value is about 3. $\hat{\sigma}_{v_i}^2$ has a range of 0.6 to 9.3, with a typical value of 1.5. The resulting θ_i's range between 0.35 and 0.68, but most are about 0.6.

With the above estimates from the first stage, generalized least squares is performed according to Eqns. (6.24) and (6.25), producing the estimated coefficients and other statistics which are reported in Exhibit 6.2 in the main chapter. We compared these results to those of three other estimators. (In order to facilitate the comparison of the results, we forced each of these alternative estimators to use the same selection of variables as the PHEC procedure.) The other estimators are presumably less efficient than the PHEC procedure. They are discussed merely to indicate to what extent the estimated coefficients are robust with respect to varying estimators, and to demonstrate how estimated standard errors and other statistics can be affected by the application of naive models. The alternative estimators are as follows:

(1) We applied ordinary least squares to the pooled raw data ("POLS"). This procedure, of course, would be the generalized least squares estimator only under the assumptions of homoscedasticity and independence of the errors.

(2) We ran generalized least squares on the pooled data, recognizing the heteroscedasticity but not the correlation across time ("PH"). We weighted each observation by $(\hat{\sigma}_{u_i}^2 + \hat{\sigma}_{v_i}^2)^{1/2}$, where the two variances for each customer were identical to those used in the PHEC procedure.

(3) In a two-stage procedure, we assumed that there were two error components, but that both were homoscedastic ("PEC"). The estimated residuals from the PHEC first stage were

$$\hat{w}_{it} = y_{it} - x_{it}\hat{\beta} . \qquad (6.A.3)$$

Regression Analysis of Local and Suburban Usage 163

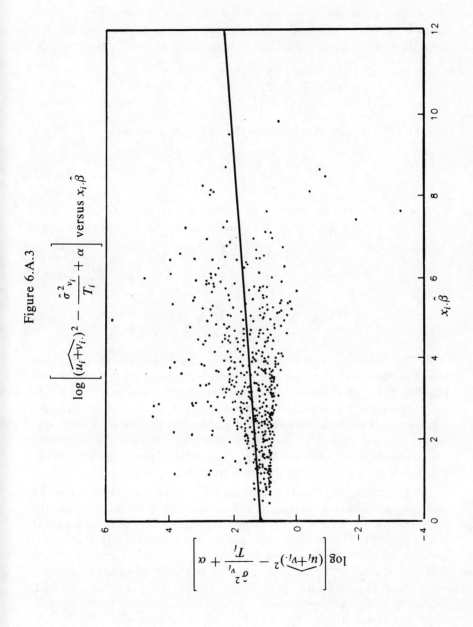

Figure 6.A.3 $\left[\log \widehat{(u_i+v_i.)^2} - \dfrac{\hat{\sigma}^2_{v_i}}{T_i} + \alpha \right]$ versus $x_i.\hat{\beta}$

Using these residuals, we estimated the variances of the two error components as follows: using the set S of customers who are observed for more than one month, the estimated variance of the monthly error was

$$\hat{\sigma}_v^2 = \frac{1}{\sum_{i \in S} T_i} \sum_{i \in S} \sum_{t=1}^{T_i} \frac{T_i}{T_i - 1} (\hat{w}_{it} - \hat{w}_{i\cdot})^2 \ ; \qquad (6.A.4)$$

and the estimated variance of the customer error was

$$\hat{\sigma}_u^2 = \frac{1}{N_S} \sum_{i \in S} \left[\hat{w}_{i\cdot}^2 - \frac{\hat{\sigma}_v^2}{T_i} \right], \qquad (6.A.5)$$

where N_S is the number of customers in set S. In this case, $\hat{\sigma}_v^2 = 1.606$ and $\hat{\sigma}_u^2 = 5.032$. Then we ran this regression:

$$y_{it} - \hat{\theta}_i y_{i\cdot} = [x_{it} - \hat{\theta}_i x_{i\cdot}]\beta + w_{it} \ , \qquad (6.A.6)$$

where

$$\hat{\theta}_i \equiv 1 - \left(\frac{\hat{\sigma}_v^2}{\hat{\sigma}_v^2 + T_i \hat{\sigma}_u^2} \right)^{1/2}. \qquad (6.A.7)$$

The outcome of the comparison of estimators was that the estimates of the F, R^2, and standard errors were dramatically affected by whether or not an estimator recognized the correlations of errors across months, although not particularly by whether it recognized the heteroscedasticity. The coefficients were affected only moderately by the stochastic assumptions. The only coefficients which differed by more than one standard error (as estimated by the PHEC procedure) from the PHEC estimate were two on the time index for 1972.

To test the hypothesis of constant coefficients across months, we also ran the following regression: We used exactly the same estimated variances for the two separate error components for each customer as in the full PHEC procedure, but we did not constrain the coefficients on each variable in the second stage to be the same across the three different months ("HEC"). Comparing the sum of squared residuals from the PHEC and HEC procedures yielded an $F(76, 1426) = 1.002$, indicating that the constraint of equality of coefficients across months could not be rejected at any reasonable significance level.

CHAPTER 7

The Effects of Time of Day on Local and Suburban Calling Frequencies and Conversation Times

Belinda B. Brandon, Paul S. Brandon, and Elsa M. Ancmon

7.1 Introduction

This chapter contains analyses of residential calling patterns by time of day, for weekdays and for weekends. Its emphasis is on how certain demographic household characteristics affect non-toll temporal calling patterns.

Currently, neither local nor suburban calls from Chicago have charges according to their time of day. But there is good reason for charging by time of day and thus for studying temporal calling patterns. If the network is at capacity during any moment of the day, then an additional call at that moment will cause the company to make additional investment or suffer degradation of service. If this call, on the other hand, could be shifted through a change in prices to a time when the network is not fully utilized, then no additional investment is needed. Further, during the peak period it is economically desirable to eliminate a call whose value to the customers involved falls below the cost of providing the capacity to handle the call. Similarly, in general it is socially desirable to limit the instances in which a customer is discouraged from making even a low-valued call during off-peak hours if the capacity cost of handling the call is lower yet. Charging higher prices during peak periods than during off-peak periods should reduce peak calling relative to off-peak calling.

If one envisions instituting time-of-day differentials in prices of local calls, one may be interested in the financial impact of such differentials on various demographic groups. Regulatory commissions certainly have evidenced such concerns in the past. The data that are summarized here would enable one to estimate the approximate impacts.

Another conceivable application for studying how demographics affect temporal calling patterns is traffic engineering. Traffic engineers develop forecasts of peak usage to aid their equipment planning. If past changes in usage are a guide to future changes, then at the level of aggregation of the Central Office Area, traffic engineers probably have all the data on current usage that they need for design decisions regarding Central Offices, trunks, and networks. But if some area is anticipated to change its demographic composition where it had been stable before, then the kind of information presented here may be useful as an aid to forecasting. Furthermore, there is now an increasing use made of the traffic concentrator, an intermediate switching point between the customer's location and the wire center. In order to design the economical placement of such machines, one would like to know the peak usage of the small neighborhoods that they are to serve. A special usage study for every such neighborhood might be expensive compared to estimating peak usage from the demographic information that the Census provides. In addition, a special usage study is impossible when telephone wiring must be in place before a new real estate development is completed. If an engineer can anticipate the demographic composition of those who will move into a new development, he can get an estimate of peak usage. Whether this procedure is useful depends upon the size of the errors of such estimates. Determining their size requires more study.

In this chapter, the relationship between measures of non-toll calling and demographic variables is analyzed first by rate periods and then by hour of the day. Hour-of-day results from the entire sample are reported graphically for weekdays and for weekends; hourly findings for individual demographic variables are displayed for weekdays; and weekend results are reported numerically for each demographic group by aggregating over hours. For local calls, the measures of calling for which results are reported are the number of calls, average duration of calls, and total conversation time. Results for suburban calls are presented for the entire sample for the number of calls, average duration of calls, total conversation time, and message units; but results by demographic variables are reported only for suburban message units. The rate period results are presented by the use of notched box plots. Results by hour of the day are displayed by the use of graphs of means with error bars. Only prominent highlights of each figure are discussed; the reader is free to study the figures in Appendix 7.B for additional detail.

This chapter is divided into five sections. Following this introductory section, the data that are utilized are described. The third section presents the results for rate periods. Section four presents the hour-of-day results. The last section discusses some suggestions for future research.

7.2 The Data

Only the information from the questionnaire on family income, on race, and on the age of the head of household are utilized for this chapter. It also draws upon only a portion of the data on telephone usage — information on the duration, time of day, and called prefix of every out-going local and suburban call for the sampled customers (the called prefix is used to classify the call into local or suburban).[1] These data were collected during the billing periods that began in April, 1974. The 1974 data were chosen for detailed analysis of calling patterns by time of day because for the first time complete data on the time of day for both local and suburban calls were available for the sample.

7.3 Results by Rate Periods

This section of the chapter is divided into five sub-sections. Sub-section 7.3.1 presents the results for the number of local calls, 7.3.2 for average duration of local calls, and 7.3.3 for total local conversation time. Sub-section 7.3.4 displays the results for suburban usage. The results for local and suburban message units are summarized in 7.3.5. Each of these five sub-sections presents results for the total sample; for both the black and white racial groups; for three income groups (under $5,000, $5,000 to $14,999, and $15,000 or over); and two age groups (households headed by a person 65 years of age or older and those by a person under 65). The division between age groups is chosen to be 65 years because much interest has been expressed over what the impact of price changes would be on those of age 65 or over. The figures to which the text below refers may be found in Appendix 7.B.

---

[1]Calls to directory assistance are excluded from the analysis.

The four rate periods utilized in this section are the ones in use for interstate toll rates (although in the toll rate schedule the same prices are charged during the night rate period as during the weekend rate period). The "weekday" rate period, the one normally considered to be the Bell System's peak, is from 8 a.m. to 5 p.m. Monday through Friday. The "evening" rate period is from 5 p.m. to 11 p.m. Sunday through Friday. The "night" rate period is from 11 p.m. to 8 a.m. Sunday through Friday. Finally, the "weekend" rate period is from 11 p.m. Friday through 5 p.m. Sunday. Each of the customers' local and suburban calls was assigned to the rate period during which the call originated. A difficulty is that, over a period of a month, there are different numbers of hours falling in different rate periods. In addition, since different customers are measured during different billing periods (with different numbers of weekdays and weekends), the number of hours falling in the same rate period will vary across customers. In order to eliminate these problems, the number of calls, total conversation time, and message units are each expressed as usage per hour.

The technical method of display in this section on rate periods is the "notched box plot". For a detailed explanation of the notched box plot, see Section 5.3 of Chapter 5. Briefly, if the notch for one box does not overlap the notch in another box from an independent sample, then one can reject the hypothesis that the populations from which the samples were drawn for the one box and for the other box have the same median at approximately the 5 percent significance level. It should be pointed out that the notches are relevant only for comparisons between any two groups of households, not for comparisons between rate periods; the confidence intervals are presumably narrower yet for comparisons between rate periods for a given group because the calling rate is positively correlated across rate periods for each individual.

As in Chapter 5, for the purpose of computing notches on the box plots, the data are transformed by adding the constant 1/6 and by taking the logarithm. Once the confidence intervals for the transformed data are found, the end points of the intervals are retransformed into the raw scale so that they may be entered on the box plots of the raw data.

These notches are useful because they aid the eye in judging whether a difference is meaningful or not, but, because they are approximations, the result of the Wilcoxon-Mann-Whitney rank sum test is reported in Appendix 7.A for every pair of groups. Statements in the text about significant differences rely solely upon these rank sum tests, using the standard of the 5 percent significance level.

7.3.1 Numbers of Local Calls by Rate Periods

Figures 7.B.1 through 7.B.4 examine the relationship between the number of local calls and demographic characteristics by rate periods. All of the figures graph the number of calls per hour on the vertical axes; rate periods are on the horizontal axes, and, for most figures, there are several demographic groups for each rate period. The results for the entire sample are displayed in Figure 7.B.1. The greatest median calling rate for this sample is for the "evening" rate period, whereas the peak for Chicago as a whole is known to be the "weekday" rate period. This difference is not surprising, since our sample consists only of residence customers and since the peak for Chicago as a whole is determined by business and residence calling combined.

Figure 7.B.2, for the two principal racial groups, shows that the rate period with the highest median calls per hour for black customers is the evening, whereas for white customers there is little difference between the median calling rate for the weekday and evening rate periods. The median number of calls for every one of the four rate periods is significantly greater for black customers than for white customers.

Figure 7.B.3 portrays the results for three family income groups: under $5,000, $5,000 to $14,999, and $15,000 or over per year. The number of calls tends to be higher for the group with incomes of $15,000 or more, but none of the differences is significant. The observant reader may note that, for both the weekday and weekend rate periods, the median for every income group in Figure 7.B.3 exceeds the median for the entire sample in Figure 7.B.1. The reason is that those who did not complete the questionnaire have a significantly lower calling rate than those who did (see Chapter 3).

As shown in Figure 7.B.4, households that are headed by persons of age 65 or over make considerably fewer calls in the system's off-peak rate periods — evening, night, and weekend — than other households do; these differences are statistically significant. The peak for the households with older heads is the weekday rate period, when their calling rate is not significantly different from the others. (The senior citizens' box plot for the night rate period is collapsed because the lower quartile is coincident with the median; *i.e.*, at least half of them made no calls in that rate period.)

7.3.2 Average Durations of Local Calls by Rate Periods

Recall from the discussion in Section 5.4 of Chapter 5 above that when durations of calls are recorded, a rounding bias is imparted to the durations. Here the estimated bias is removed from each call duration before the aggregation into rate periods, by the same procedure as in Chapter 5.

Figures 7.B.5 through 7.B.8 display the findings by rate period for customers' average durations of local calls. Summary statistics of the distribution of customers' average durations are graphed on the vertical axes; rate periods and demographic groups are on the horizontal axes. First refer to Figure 7.B.5. The median of the customers' average durations of local calls is highest for the evening rate period (5.6 minutes), next highest for the weekday and weekend rate periods (3.9 minutes in both cases), and lowest at night (1.5 minutes). A pattern similar to this holds for every demographic group examined in this sub-section except for the households with older heads (see below).

Figure 7.B.6 displays the results by race. For every rate period, the median for black customers is significantly higher than that for white customers.

The next figure — number 7.B.7 — contains the results for income groups. For every rate period the median of the average durations of local calls is higher for the lowest-income group than for the middle-income group; in turn they are higher for the middle-income group than for the highest-income group. These differences are statistically significant between the lowest and highest income groups for the weekday and evening rate periods, and between the lowest and middle income groups for the weekday rate periods. In addition, there are significant differences at night between the lowest-income group and both the middle- and highest-income groups.

Figure 7.B.8, the last in this sub-section, graphs the relationship between average durations of local calls and rate periods for households whose heads are at least 65 years of age and for those whose heads are under 65. For the older group, the largest median occurs for calls placed during the weekend, then weekday, then evening, then night. The only significant difference between the age groups is for weekends, when the older group has longer calls.

7.3.3 Total Local Conversation Times by Rate Periods

Total local conversation times are displayed in Figures 7.B.9 through 7.B.12. Total local conversation minutes per hour are graphed on the vertical axes and rate periods with demographic groups on the horizontal axes. All calculations use the asymptotically unbiased data set employed in the previous sub-section. The qualitative results are similar to those for the number of local calls. Figure 7.B.9 presents the results for the entire sample. The peak period is the evening, followed by weekday, weekend, and then night. This same pattern is qualitatively followed in Figures 7.B.10 and 7.B.11, but not in Figure 7.B.12 (for senior citizens).

Figure 7.B.10 displays the results for black and white customers. Black customers have a significantly higher median of total local conversation times for every rate period than the white customers do, but the difference is less pronounced for weekdays than for the other rate periods.

Now compare the results for the three income groups in Figure 7.B.11. The lowest-income group has the highest median of total local conversation times for each rate period, but there is no statistically significant difference.

For households headed by a person 65 years of age or older, Figure 7.B.12 shows that the highest median of total local conversation times occurs in the weekday rate period, followed by evening, weekend, then night. Older households have a significantly lower local usage than the others in the evening, night, and weekend rate periods.

7.3.4 Suburban Calling by Rate Periods

This sub-section presents the results for the number of suburban calls, average duration of suburban calls, total suburban conversation time, and number of suburban message units for the entire sample; in addition, results for the two racial groups, the three income groups, and two age groups are presented for the number of suburban message units, which is a useful summary measure.

Figure 7.B.13 graphs the numbers of suburban calls on the vertical axis versus rate periods on the horizontal axis. The peak rate period is the evening, then weekend, then (close behind) weekday, then night. (For the entire sample, the medians of all suburban usage measures are zero for night except in the case of average duration.) The reader should recall that for local calls the order was evening, weekday, weekend, then night. The rate of suburban calling is only around 5 percent of the rate of local calling.

Average durations of suburban calls (with bias removed) by rate periods are displayed in Figure 7.B.14. The peak median of average durations is in the evening rate period, with shorter durations for weekday, then weekend, then night. As the reader should recall, the median of average durations of local calls was the highest in the evening rate period, then almost tied between weekday and weekend, then considerably lower at night. The medians of average durations of suburban calls are lower than those of local calls for the evening and the weekend rate periods, slightly higher for the weekdays, and equal for nights.

As indicated in Figure 7.B.15, the median of total suburban conversation times is greatest in the evening rate period, then considerably lower in weekday, then weekend, then night. This pattern is qualitatively the same as was found for total local conversation times.

Figures 7.B.16 through 7.B.19 graph suburban message units on the vertical axes and rate periods with demographic groups on the horizontal axes. In Figure 7.B.16, for the entire sample, we see a similar pattern to what we saw for total suburban conversation time. Figure 7.B.17 exhibits results for the two races. The median number of suburban message units is significantly higher for white customers than for black customers in the weekday, evening, and weekend rate periods. No significant difference is found for night.

As one compares the three income groups in Figure 7.B.18, one notices that the median number of suburban message units increases as the level of income increases, for every rate period except night. There is no significant difference at night. But during each of the other rate periods, each income group is significantly different from the other two except for one instance: the two lowest income groups are insignificantly different during weekdays.

The last figure in the sub-section, Figure 7.B.19, shows the number of suburban message units used by the households headed by persons of different ages. Senior citizens' medians of suburban message units follow a similar pattern for that seen for the other age group, with the peak rate period's being the evening. This group's suburban calling pattern differs from its local calling pattern in that its peak for local calling is the weekday rate period. There is no significant difference between the age groups' suburban usages.

7.3.5 Local and Suburban Message Units

This sub-section graphs the results for total local and suburban message units by rate periods. The rate period with the highest median message units per hour is the evening, followed by day, then weekend, then night (see Figure 7.B.20).

In Figure 7.B.21 black customers have a significantly higher median usage of total message units in the night and weekend rate periods than do white customers.

In Figure 7.B.22, the highest-income group has significantly higher message-unit usage than the middle-income group during evenings, and has significantly higher usage than the lowest-income group in both the evening and weekend rate periods.

In the last figure in this sub-section, Figure 7.B.23, the 65-or-over group has a significantly lower median number of message units than the younger group in all rate periods except weekday.

7.4 Results by Hours of the Day

This section is divided into five sub-sections. The first three of these sub-sections cover the number of local calls, average duration of local calls, and total local conversation time. Sub-section 7.4.4 deals with calls to the suburbs — their number, durations, total conversation time, and message units. The last sub-section presents the results for total local and suburban message units. While the method of display in the previous section was the notched box plot, this section utilizes plots of means with error bars. The large proportion of households who make no calls during a particular hour of the day over the course of a month creates too great a problem for box plots to handle well. For instance, during the entire period comprised of the 20-odd hours between 10-11 a.m. on the weekdays in the month of study, no local calls are made by 30 percent of the sample, and no suburban calls are made by 78 percent.

Plots for the entire sample have 95 percent confidence intervals around the means — 1.96 standard errors. But for any demographic characteristic, as a rough guide to significance, the size of the error bars are chosen to be 1.7 standard errors around the mean. As in the case of the notches on the box plots of the last section, 1.7 is a compromise between the extremes of 1.39, which would be appropriate if one were comparing two groups for whom the variances of their means were equal, and near 1.96, if one were comparing two groups with very different variances. If one is interested in comparing two groups for a particular hour, then one can see whether the error bars for the two groups overlap. If they do not, then one can reject the hypothesis that the two groups have the same mean at that hour at approximately the 5 percent significance level.

One may observe that, because there is so much noise relative to the amount of data, the number of instances are small in which one can reject the hypothesis that one group has the same mean as another at some hour; *i.e.*, the power of the test implicit in the displays in this section is small. It is hoped that the sacrifice in power in this section is compensated by the greater detail that is provided by estimates for each hour of the day. The aggregation over many hours into the rate periods of the previous section provides greater power to test for group differences. In fact, if there is a consistent pattern across hours, then more credence should be placed on an estimated difference between groups than is indicated by the tests for individual hours. The reader should also keep in mind that the issue of the multiple comparison problem discussed in Chapter 1 applies with considerable force in this section: not much importance should be attached to a significant difference for a single hour if it is not reinforced by a consistent pattern for nearby hours.

A warning in the previous section should be repeated here; do not use the plotted error bars for tests between hours rather than between groups. The positive correlation of usage among hours implies that two hours' error bars could be overlapping, yet the hours could still be significantly different.

The reader may note that, when more than one group is shown on a plot, the points of one group are slightly offset horizontally from the other(s). The purpose is merely to improve the visibility of the error bars. Also, the hours of midnight to 6 a.m. are aggregated into one point so as to reduce the crowding of the other points, since it is presumed that there is little interest in the detail of calling patterns in the early morning hours.

7.4.1 Numbers of Local Calls by Hours of the Day

Figures 7.B.24 through 7.B.29 graph the numbers of local calls on the vertical axes and the hours of the day on the horizontal axes. Figure 7.B.24 presents the results for the entire sample for weekdays. The mean number of local calls rises rapidly to 9-10 a.m. and then remains relatively stable until 3-4 p.m., when it jumps up and stays at a high level for most of the next several hours. The peak of 0.29 calls per hour occurs at 6-7 p.m. The second highest calling rate is 0.26 at both 5-6 p.m. and 7-8 p.m.

Figure 7.B.25 graphs the entire sample for weekends. The number of local calls rises rapidly to a peak of 0.26 calls per hour at 10-11 a.m., then drifts downward until 7-8 p.m., after which the local calling rate begins plunging.

The weekday number of calls is displayed for each hour by race in Figure 7.B.26. The local calling rate is higher for black customers than for white customers for every weekday hour, especially in the evening. The peak calling rate for black customers is pronounced at 6-7 p.m., while for white customers the hours of 3-4 p.m. and 6-7 p.m. form a twin peak. No results are graphed by demographic groups for any usage measure for weekends, because it is presumed that there is less interest in the detailed calling patterns for this off-peak period. It may be useful, however, to state that over the weekend as a whole the mean number of calls per hour for white customers is half that of black customers, and the difference is statistically significant. The aggregate weekend usages and aggregate weekday usages can be seen in Appendix 7.C, which lists the mean value of each usage measure for each demographic group.

Figure 7.B.27 presents the weekday results for the numbers of local calls for income groups. There is no significant difference between income groups for any weekday hour. Even the ranking of groups jumps from hour to hour except in the evening hours after 6 p.m. In the latter period, the higher the income, the higher the calling rate. As shown in Appendix 7.C, the mean number of calls per hour aggregated over the entire weekend is not significantly different between income groups.

Figure 7.B.28 is a comparison of the calling pattern of households whose heads are under 65 years of age with that of households whose heads are 65 years of age or over. There is a prominent time-of-day difference between the two groups. The 65-or-over group has its peak between 9 a.m. and 12 noon, when its calling rate is higher than that of the other group, while the under-65 group peaks at 6-7 p.m. at a higher level than the older group. The weekend calling rate is significantly lower for the older group.

Figure 7.B.29 is inserted as a digression. Households with at least one child of age 18 or under are plotted by a circle, and households without any children 18 or under are plotted by a triangle. Households with children have a jump from 0.26 to 0.38 local calls per hour (a 46 percent increase) from 2-3 p.m. to 3-4 p.m.; but households without children only have an increase from 0.13 to 0.18 for the corresponding period (a 38 percent rise). The authors speculate that this difference between groups is attributable to children's returning home from school. From this information alone, one does not know whether children themselves call or whether parents tend to arrive home at that time and call.

7.4.2 Mean Durations of Local Calls by Hours of the Day

Figures 7.B.30 through 7.B.34 graph the means of households' average durations of graphed on the vertical axes and hours of the day on the horizontal axes. As in the previous section, on rate periods, this section removes the estimated bias from durations.

A note of caution is that the estimated standard errors may somewhat understate the true standard errors because of the existence of heteroscedasticity. Different households make different numbers of calls, so their individual means are estimated with varying accuracies. It would have been possible to weight each household's average duration by $1/(s^2 + t^2/n_i)$, where s^2 is an estimate of the variance across households' expected durations, t^2 is an estimate of the variance across the durations of calls for an individual household, and n_i is the number of calls made by the ith household. However, the average duration for a household is correlated with its number of calls, so this weighted estimator would be biased and inconsistent. The sample mean is used here in preference to a biased estimator and in preference to developing a considerably more complex model that might avoid this bias while removing the heteroscedasticity.

Figures 7.B.30 and 7.B.31 plot the mean durations of local calls for weekdays and for weekends, respectively, for the entire sample. During weekdays a peak of 9.2 minutes per call occurs at 8-9 p.m., with other evening hours' also having high mean durations. Daytime calls average 5 to 6 minutes. Early morning calls tend to be short. The mean durations for weekends rise gently to 7 p.m., then rise faster to a peak of 8.4 minutes at 10-11 p.m.

In Figure 7.B.32, black customers have higher mean durations than do white customers for all but one hour, although there is a statistically significant difference for only a few individual hours. According to Appendix 7.C, black customers have a significantly higher mean duration for weekends than do white customers.

Figure 7.B.33 displays mean durations by income. Patterns across hours are qualitatively similar for all the income groups, except that the lower the income, the earlier the peak. There are few significant differences for individual hours, but the ranking of groups is fairly consistent: the mean duration is higher the lower the income is. The test in Appendix 7.C for weekdays as a whole and for weekends as a whole confirms this tendency; and the difference between the highest- and lowest-income groups is statistically significant.

No systematic difference appears in Figure 7.B.34 between households headed by a person 65 years of age or over and households headed by a person under 65 years of age. While the mean durations are not significantly different between the age groups during weekends, the estimate for the older group is appreciably larger than for the younger.

7.4.3 Total Local Conversation Times by Hours of the Day

Figures 7.B.35 through 7.B.39 present the results for total local conversation time. Total conversation minutes are graphed on the vertical axes and the hours of the day on the horizontal axes. As is done with average durations, the bias due to rounding is removed. Refer to Figure 7.B.35 for total local conversation times for weekdays for the entire sample. This measure displays a broad peak of 2.0 minutes per hour, for both 7-8 p.m. and 8-9 p.m. Weekend total local conversation times, as shown in Figure 7.B.36, show very little variation from 10 a.m. to 10 p.m.: for every hour in that interval mean usage lies within the range of 1.05 to 1.30. The reader should recall that there was a downward drift in the number of local calls from 10 a.m. to 10 p.m.; evidently, that pattern is compensated by an upward drift in the mean durations.

In Figure 7.B.37 the hour of the peak in total local conversation times is the same for the two racial groups — 8-9 p.m. — whereas, as shown in Section 7.4.1, the white group has a peak in the numbers of local calls at 3-4 p.m. But according to the measure, total local conversation time, the black group has much greater usage in the evening relative to daytime usage than does the white group; the level of the evening peak is also twice that of the white group. As shown in Appendix 7.C, black customers have a significantly and dramatically higher mean of total local conversation times on weekends than do white customers.

Total local conversation times by income, as shown in Figure 7.B.38, are difficult to disentangle. However, the under-$5,000 group has the highest usage for most of the day-time hours. The qualitative pattern across hours is similar among the income groups. The weekend usage of the income groups is ranked inversely to income, but only the difference between the lowest and highest income groups is significantly different.

As displayed in Figure 7.B.39, households headed by persons of age 65 or over have a peak of total local conversation times in the evening — as all other groups do — whereas their *numbers* of local calls have a morning peak. Still, the difference between morning and evening total local conversation times is less pronounced for them than it is for households with younger heads. Weekend usage is estimated to be higher for the households with younger heads than for the others, but the difference is not significant.

7.4.4 Suburban Calling by Hours of the Day

This sub-section, utilizing Figures 7.B.40 to 7.B.50, describes the pattern across hours of the day of the following measures of calling to the suburbs by the entire sample of Chicago customers: the number of calls, average duration of calls, total conversation time, and message units. In addition, it reports the pattern of suburban message units for the same demographic groups as in the previous sections.

Figures 7.B.40 and 7.B.41 present the results for the entire sample for the numbers of suburban calls for weekdays and for weekends, respectively. The pattern of the numbers of calls to the suburbs is qualitatively similar to the pattern of calls within Chicago both for weekdays and for weekends (*cf.* Figures 7.B.24 and 7.B.25), but the number of suburban calls is an order of magnitude smaller than that of local calls. The mean durations of suburban calls is portrayed in Figures 7.B.42 and 7.B.43, with estimated bias removed. In both figures, because of the large amount of noise in the data, the pattern of average durations of suburban calls is very irregular across hours of the day. The peak seems to be at 9-10 p.m. both for weekdays and weekends, when the mean durations are 10.1 and 8.5 minutes per call, respectively. In Figure 7.B.44, the peak of asymptotically unbiased total suburban conversation times for weekdays occurs at 7-8 p.m. (but suburban usage is still relatively higher in the afternoon than is the case for local calling). There is a prominent peak at the hour 11-12 noon in Figure 7.B.45 for weekends. Recall that total *local* conversation times were very level during weekend days.

The peak for suburban message units for weekdays, as shown in Figure 7.B.46, is at 7-8 p.m. Figure 7.B.47 shows a prominent peak in suburban message units at 11-12 a.m. on weekends, as was the case for total suburban conversation times.

For most hours of the day, as shown in Figure 7.B.48, white customers have substantially and statistically significantly greater suburban message unit usage than do black customers. The peak for white customers is at 8-9 p.m., while the peak for black customers is at 12-1 p.m. Weekend suburban calling for white customers is also significantly higher than for black customers.

In Figure 7.B.49, for the majority of hours the income groups are ranked as one might expect — the higher the income, the higher the suburban message unit usage. The same is true for weekends; the difference between the highest and lowest income groups is statistically significant. One sees in Figure 7.B.50 that message unit usage tends to be lower for households whose heads are 65 years or older, but in general statistically not significantly so. The difference between groups is most pronounced in the evening. Weekend usage is insignificantly higher for the younger group.

7.4.5 Total Local and Suburban Message Units by Hours of the Day

Figures 7.B.51 and 7.B.52 present the results for total local and suburban message units for weekdays and weekends, respectively. The weekday message unit usage is high between 3 p.m. and 9 p.m., with a peak at 6-7 p.m. However, on weekends the peak message unit usage occurs at the hour 11 a.m. to 12 noon.

In Figure 7.B.53, when one examines the number of message units by race, one can see that the white customers have a higher message unit usage except for late night and early morning. Their peak occurs at 4-5 p.m., whereas the black customers' peak occurs at 6-7 p.m. As shown in Appendix 7.C, the sample of white customers has insignificantly higher usage during weekdays as a whole than does the sample of black customers; on the other hand, the black customers have insignificantly higher aggregate weekend usage than do the white customers. The reader should recall that black customers have a higher number of local calls for every hour of the day, and white customers have a higher number of suburban message units.

In Figure 7.B.54 the results of total message units by income groups are presented. The highest income group has the greatest message unit usage for most of the day; the exceptions are around lunch and in the interval 12 midnight to 7 a.m. The $15,000-or-over group peaks at 4-5 p.m. The $5,000-to-14,999 group peaks at 3-4 p.m., closely followed by 4-5 p.m. and 7-8 p.m. The peak usage of the lowest-income group occurs at 6-7 p.m., nearly matched by 3-4 p.m. In the cases of both aggregate weekday and weekend usage, the higher the income is, the higher is the mean of total message units, although a statistically significant difference is observed only between the lowest- and highest-income groups for weekday usage.

In the last figure, one sees that the households with younger heads have the higher usage for the whole day except between 9 a.m.-12 noon; they also have a more prominent evening peak than the households with older heads. The aggregate weekday and weekend message unit usages are both significantly higher for the households with younger heads.

7.5 Conclusions and Suggestions for Future Research

One may briefly summarize the results on local usage as follows: households with heads who are white or over 65 years of age make a larger fraction of their calls during the day on weekdays than do the others; but income appears to make little difference in usage across hours of the day. In contrast, it is the black and lowest-income households that make a larger fraction of their suburban calls during the day on weekdays than the others; and age seems to affect the suburban time-of-day pattern very little. As for total local and suburban message units, white customers have a somewhat higher usage during the day on weekdays than do black customers, but the black customers have higher usage at night and on weekends than do white customers; message-unit usage rises with income, but the relative patterns across times of day are only mildly affected by income. Finally, the message-unit usage of the older households is heavily concentrated during the day on weekdays.

This chapter has shown a selection of the kind of results that one can obtain by associating demographic characteristics with usage by time of day. Much more could be done. The scope of this chapter has been limited to the analysis of time-of-day and rate period usage patterns by one demographic characteristic at a time. A natural extension is to a simultaneous multiple-characteristic analysis. For instance a regression could be run of each measure of usage in each hour against a set of binary variables representing the divisions of several demographic characteristics.

A more sophisticated estimation procedure may yield estimates of the mean number of calls, average duration, and total conversation time that have lower variances than those calculated herein. A suggestion by Kenneth Wachter utilizes the additional information of the negative correlation between the number of calls and average durations. The fact that a household makes a call of given non-zero duration reduces the time it has remaining in an hour in which to make another call; and the longer the one call, the less time is available to make another. One might treat call durations as exponentially distributed with one parameter and the waiting time between calls as exponentially distributed with another parameter. (In addition, there might be a probability of less than one of what one might interpret as "being home" to activate the finite expected waiting time.) One can then derive the expected number of calls and the expected duration jointly, along with their variances. The expected total conversation time and its variance might also fall out of such a procedure.

Appendix 7.A

Rank Sum Tests for Differences Between Demographic Groups by Rate Periods

Probability of Observing a Rank Sum No Larger than Observed for the Smaller Sample

Pairs of Groups			Local			Suburban	Local and Suburban
			Number of Calls	Average Duration	Total Conv. Time	Message Units	Total Message Units
Smaller	Larger	Rate Periods			Probabilities		
Black	White	Weekday	**0.999**	**0.985**	**0.999**	**0.000**	0.487
		Evening	**0.999**	**0.999**	**0.999**	**0.000**	0.967
		Night	**0.999**	**0.999**	**0.999**	0.516	**0.999**
		Weekend	**0.999**	**0.999**	**0.999**	**0.000**	**0.999**
<$5,000	$5,000-14,999	Weekday	0.736	**0.998**	0.969	0.246	0.676
		Evening	0.179	0.937	0.532	**0.002**	0.071
		Night	0.642	**0.980**	0.757	0.427	0.577
		Weekend	0.527	0.964	0.893	**0.010**	0.115
$5,000	≥$15,000	Weekday	0.421	**0.995**	0.925	**0.005**	0.188
		Evening	0.048	**0.997**	0.523	**0.000**	**0.001**
		Night	0.198	**0.999**	0.651	0.339	0.245
		Weekend	0.358	0.961	0.849	**0.000**	**0.022**
≥$15,000	$5,000-14,999	Weekday	0.799	0.444	0.546	**0.992**	0.931
		Evening	0.885	0.089	0.560	**0.999**	**0.995**
		Night	0.931	0.065	0.619	0.625	0.852
		Weekend	0.747	0.313	0.608	**0.998**	0.929
≥65 yrs.	<65 yrs.	Weekday	0.174	0.811	0.285	0.299	0.256
		Evening	**0.000**	0.184	**0.000**	0.327	**0.000**
		Night	**0.000**	0.624	**0.000**	0.295	**0.000**
		Weekend	**0.000**	0.982	**0.021**	0.096	**0.000**

* Differences that are significant at the 5 percent level or better are shown in bold type.

Appendix 7.B

Local and Suburban Usage by Demographic Group: Box Plots by Rate Period and Means by Hour of the Day

Figure Number	Title	Page
7.B.1	Number of Local Calls per Hour by Rate Periods for Entire Sample	186
7.B.2	Number of Local Calls per Hour by Rate Periods by Race	186
7.B.3	Number of Local Calls per Hour by Rate Periods by Income	187
7.B.4	Number of Local Calls per Hour by Rate Periods by Age of Head of Household	187
7.B.5	Average Durations of Local Calls by Rate Periods for Entire Sample	188
7.B.6	Average Durations of Local Calls by Rate Periods by Race	188
7.B.7	Average Durations of Local Calls by Rate Periods by Income	189
7.B.8	Average Durations of Local Calls by Rate Periods by Age of Head of Household	189
7.B.9	Total Local Conversation Time per Hour by Rate Periods for Entire Sample	190
7.B.10	Total Local Conversation Time per Hour by Rate Periods by Race	190
7.B.11	Total Local Conversation Time per Hour by Rate Periods by Income	191
7.B.12	Total Local Conversation Time per Hour by Rate Periods by Age of Head of Household	191
7.B.13	Number of Suburban Calls per Hour by Rate Periods for Entire Sample	192
7.B.14	Average Durations of Suburban Calls by Rate Periods for Entire Sample	192
7.B.15	Total Suburban Conversation Time per Hour by Rate Periods for Entire Sample	193
7.B.16	Suburban Message Units per Hour by Rate Periods for Entire Sample	194

7.B.17	Suburban Message Units per Hour by Rate Periods by Race	194
7.B.18	Suburban Message Units per Hour by Rate Periods by Income	195
7.B.19	Suburban Message Units per Hour by Rate Periods by Age of Head of Household	195
7.B.20	Total Local and Suburban Message Units per Hour by Rate Periods for Entire Sample	196
7.B.21	Total Local and Suburban Message Units per Hour by Rate Periods by Race	196
7.B.22	Total Local and Suburban Message Units per Hour by Rate Periods by Income	197
7.B.23	Total Local and Suburban Message Units per Hour by Rate Periods by Age of Head of Household	197
7.B.24	Mean Number of Local Calls per Hour by Time of Day, Weekdays, for Entire Sample	198
7.B.25	Mean Number of Local Calls per Hour by Time of Day, Weekends, for Entire Sample	198
7.B.26	Mean Number of Local Calls per Hour by Time of Day, Weekdays, by Race	199
7.B.27	Mean Number of Local Calls per Hour by Time of Day, Weekdays, by Income	199
7.B.28	Mean Number of Local Calls per Hour by Time of Day, Weekdays, by Age of Head of Household	200
7.B.29	Mean Number of Local Calls per Hour by Time of Day, Weekdays, Households with and without Children	200
7.B.30	Mean Durations of Local Calls by Time of Day, Weekdays, for Entire Sample	201
7.B.31	Mean Durations of Local Calls by Time of Day, Weekends, for Entire Sample	201
7.B.32	Mean Durations of Local Calls by Time of Day, Weekdays, by Race	202
7.B.33	Mean Durations of Local Calls by Time of Day, Weekdays, by Income	202
7.B.34	Mean Durations of Local Calls by Time of Day, Weekdays, by Age of Head of Household	203
7.B.35	Mean Total Local Conversation Time per Hour by Time of Day, Weekdays, for Entire Sample	204
7.B.36	Mean Total Local Conversation Time per Hour by Time of Day, Weekends, for Entire Sample	204
7.B.37	Mean Total Local Conversation Time per Hour by Time of Day, Weekdays, by Race	205

7.B.38	Mean Total Local Conversation Time per Hour by Time of Day, Weekdays, by Income	205
7.B.39	Mean Total Local Conversation Time per Hour by Time of Day, Weekdays, by Age of Head of Household	206
7.B.40	Mean Number of Suburban Calls per Hour by Time of Day, Weekdays, for Entire Sample	207
7.B.41	Mean Number of Suburban Calls per Hour by Time of Day, Weekends, for Entire Sample	207
7.B.42	Mean Durations of Suburban Calls by Time of Day, Weekdays, for Entire Sample	208
7.B.43	Mean Durations of Suburban Calls by Time of Day, Weekends, for Entire Sample	208
7.B.44	Mean Total Suburban Conversation Time per Hour by Time of Day, Weekdays, for Entire Sample	209
7.B.45	Mean Total Suburban Conversation Time per Hour by Time of Day, Weekends, for Entire Sample	209
7.B.46	Mean Suburban Message Units per Hour by Time of Day, Weekdays, for Entire Sample	210
7.B.47	Mean Suburban Message Units per Hour by Time of Day, Weekends, for Entire Sample	210
7.B.48	Mean Suburban Message Units per Hour by Time of Day, Weekdays, by Race	211
7.B.49	Mean Suburban Message Units per Hour by Time of Day, Weekdays, by Income	211
7.B.50	Mean Suburban Message Units per Hour by Time of Day, Weekdays, by Age of Head of Household	212
7.B.51	Mean Total Local and Suburban Message Units per Hour by Time of Day, Weekdays, for Entire Sample	213
7.B.52	Mean Total Local and Suburban Message Units per Hour by Time of Day, Weekends, for Entire Sample	213
7.B.53	Mean Total Local and Suburban Message Units per Hour by Time of Day, Weekdays, by Race	214
7.B.54	Mean Total Local and Suburban Message Units per Hour by Time of Day, Weekdays, by Income	214
7.B.55	Mean Total Local and Suburban Message Units per Hour by Time of Day, Weekdays, by Age of Head of Household	215

186 *The Effect of Demographics on Telephone Usage*

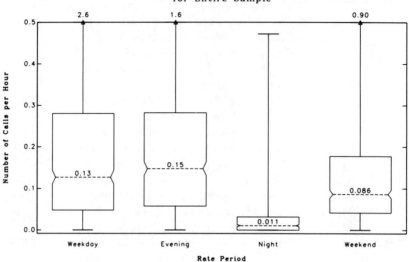

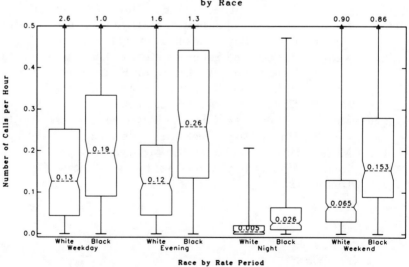

The Effects of Time of Day 187

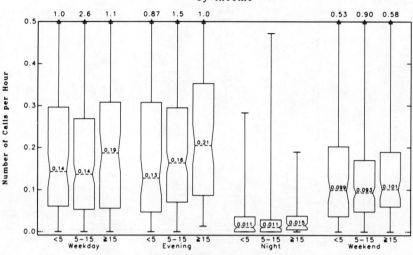

Figure 7.B.3

Number of Local Calls per Hour by Rate Periods by Income

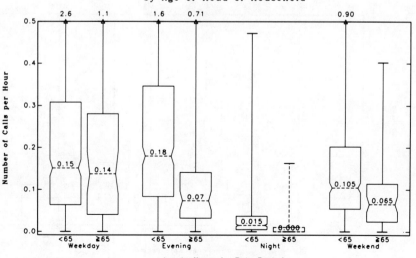

Figure 7.B.4

Number of Local Calls per Hour by Rate Periods by Age of Head of Household

188 *The Effect of Demographics on Telephone Usage*

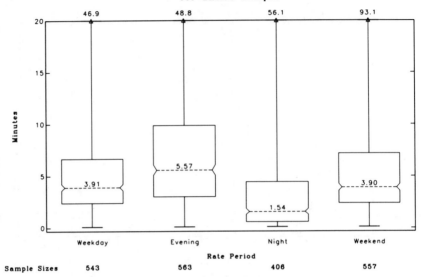

Figure 7.B.5
Average Durations of Local Calls by Rate Periods for Entire Sample

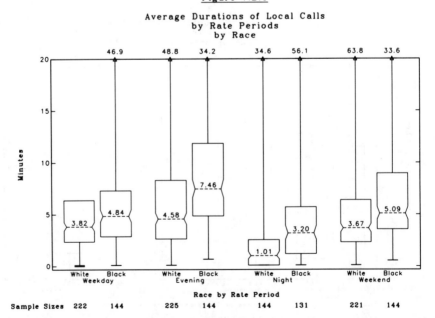

Figure 7.B.6
Average Durations of Local Calls by Rate Periods by Race

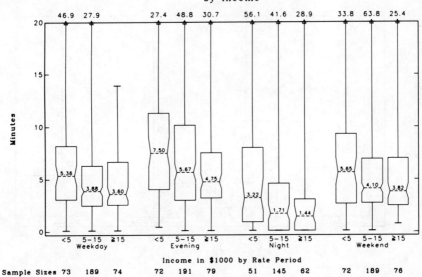

Figure 7.B.7

Average Durations of Local Calls by Rate Periods by Income

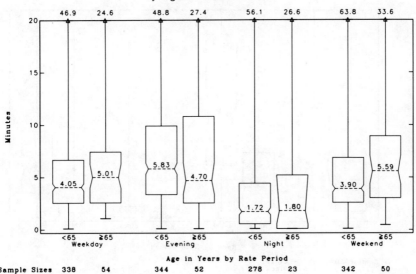

Figure 7.B.8

Average Durations of Local Calls by Rate Periods by Age of Head of Household

190 *The Effect of Demographics on Telephone Usage*

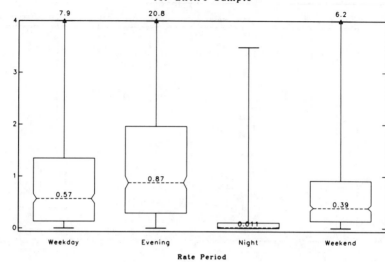

Figure 7.B.9

Total Local Conversation Time per Hour
by Rate Periods
for Entire Sample

Sample Size 573

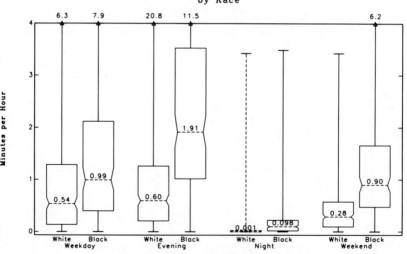

Figure 7.B.10

Total Local Conversation Time per Hour
by Rate Periods
by Race

Sample Sizes: 232 — White
146 — Black

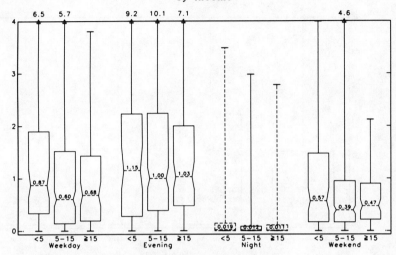

Figure 7.B.11

Total Local Conversation Time per Hour by Rate Periods by Income

Sample Sizes: 74 - Income under $5,000
              196 - Income $5,000 to 14,999
               79 - Income $15,000 or over

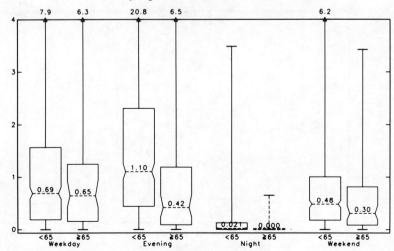

Figure 7.B.12

Total Local Conversation Time per Hour by Rate Periods by Age of Head of Household

Sample Sizes: 350 - Households with Heads Age under 65
               55 - Households with Heads Age 65 or over

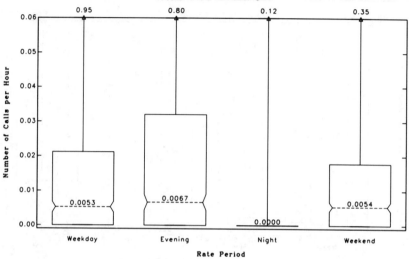

Figure 7.B.13

Number of Suburban Calls per Hour by Rate Periods for Entire Sample

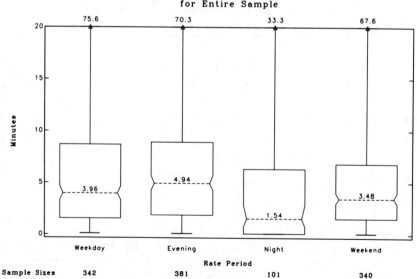

Figure 7.B.14

Average Durations of Suburban Calls by Rate Periods for Entire Sample

The Effects of Time of Day 193

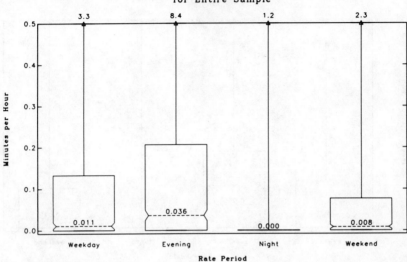

Figure 7.B.15

Total Suburban Conversation Time per Hour by Rate Periods for Entire Sample

Sample Size 573

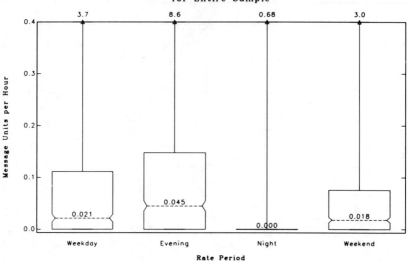

Figure 7.B.16

Suburban Message Units per Hour by Rate Periods for Entire Sample

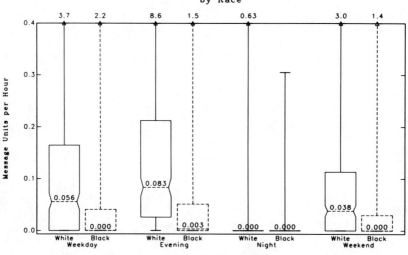

Figure 7.B.17

Suburban Message Units per Hour by Rate Periods by Race

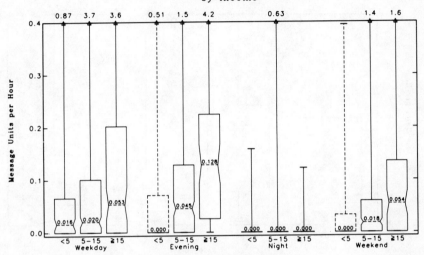

Figure 7.B.18

Suburban Message Units per Hour by Rate Periods by Income

Sample Sizes: 74 – Income under $5,000
196 – Income $5,000 to 14,999
79 – Income $15,000 or over

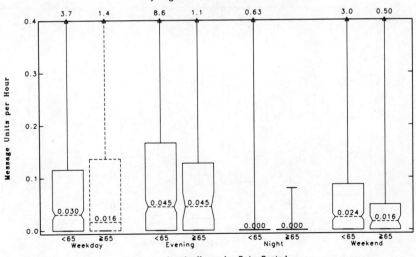

Figure 7.B.19

Suburban Message Units per Hour by Rate Periods by Age of Head of Household

Sample Sizes: 350 – Households with Heads Age under 65
55 – Households with Heads Age 65 or over

196 *The Effect of Demographics on Telephone Usage*

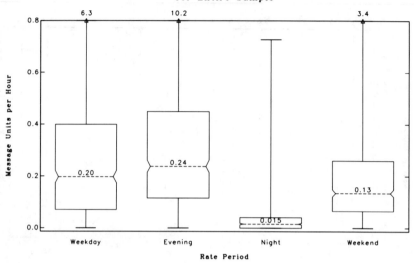

Figure 7.B.20

Total Local and Suburban Message Units per Hour by Rate Periods for Entire Sample

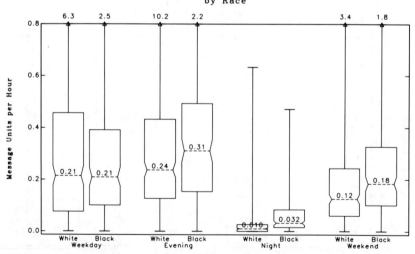

Figure 7.B.21

Total Local and Suburban Message Units per Hour by Rate Periods by Race

Figure 7.B.22

Total Local and Suburban Message Units per Hour by Rate Periods by Income

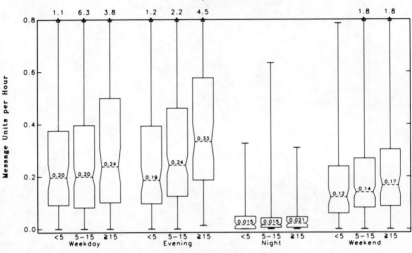

Sample Sizes: 74 – Income under $5,000
196 – Income $5,000 to 14,999
79 – Income $15,000 or over

Figure 7.B.23

Total Local and Suburban Message Units per Hour by Rate Periods by Age of Head of Household

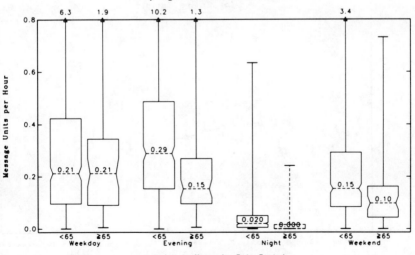

Sample Sizes: 350 – Households with Heads Age under 65
55 – Households with Heads Age 65 or over

198 *The Effect of Demographics on Telephone Usage*

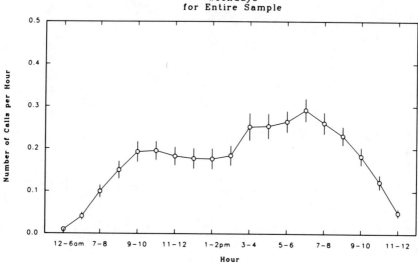

Figure 7.B.24

Mean Number of Local Calls per Hour
by Time of Day
Weekdays
for Entire Sample

Sample Size 573

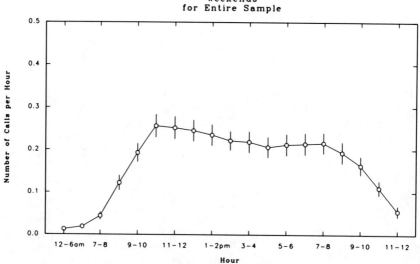

Figure 7.B.25

Mean Number of Local Calls per Hour
by Time of Day
Weekends
for Entire Sample

Sample Size 573

The Effects of Time of Day 199

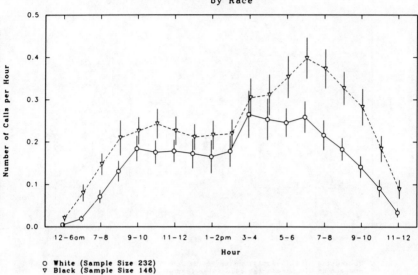

Figure 7.B.26

Mean Number of Local Calls per Hour
by Time of Day
Weekdays
by Race

○ White (Sample Size 232)
▽ Black (Sample Size 146)

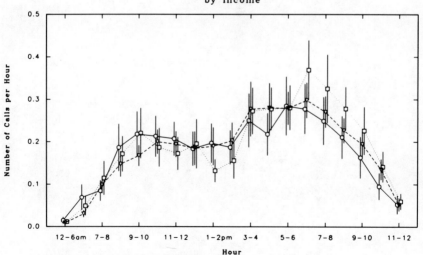

Figure 7.B.27

Mean Number of Local Calls per Hour
by Time of Day
Weekdays
by Income

○ Income under $5,000 (Sample Size 74)
▽ Income from $5,000 to $14,999 (Sample Size 196)
□ Income $15,000 or over (Sample Size 79)

200 *The Effect of Demographics on Telephone Usage*

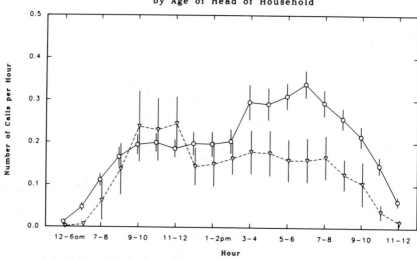

Figure 7.B.28

Mean Number of Local Calls per Hour
by Time of Day
Weekdays
by Age of Head of Household

○ Households with Heads Age under 65 (Sample Size 350)
▽ Households with Heads Age 65 or over (Sample Size 55)

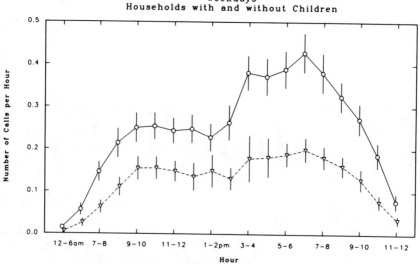

Figure 7.B.29

Mean Number of Local Calls per Hour
by Time of Day
Weekdays
Households with and without Children

○ Households with Ages 18 or under (Sample Size 200)
▽ Households without Ages 18 or under (Sample Size 220)

Figure 7.B.30
Mean Durations of Local Calls
by Time of Day
Weekdays
for Entire Sample

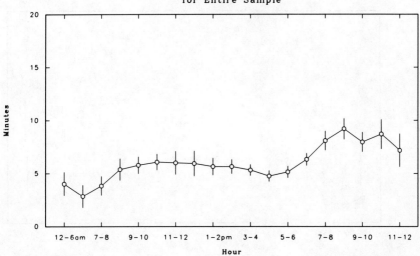

Figure 7.B.31
Mean Durations of Local Calls
by Time of Day
Weekends
for Entire Sample

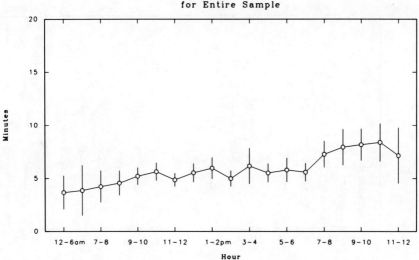

202 *The Effect of Demographics on Telephone Usage*

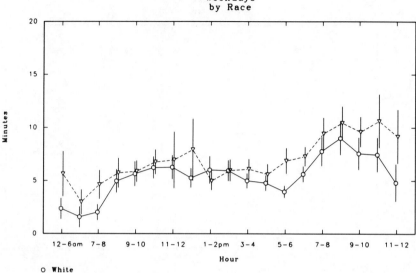

Figure 7.B.32

Mean Durations of Local Calls
by Time of Day
Weekdays
by Race

○ White
▽ Black

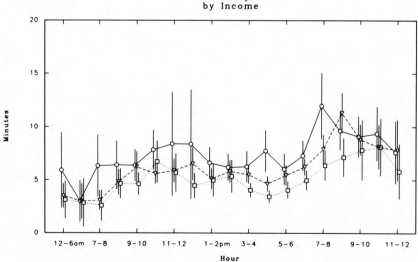

Figure 7.B.33

Mean Durations of Local Calls
by Time of Day
Weekdays
by Income

○ Income under $5,000
▽ Income from $5,000 to $14,999
□ Income $15,000 or over

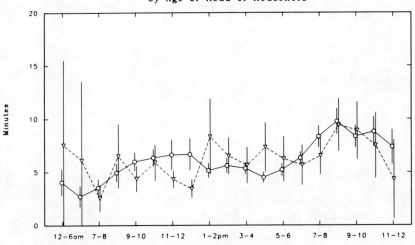

Figure 7.B.34

Mean Durations of Local Calls
by Time of Day
Weekdays
by Age of Head of Household

○ Households with Heads Age under 65
▽ Households with Heads Age 65 or over

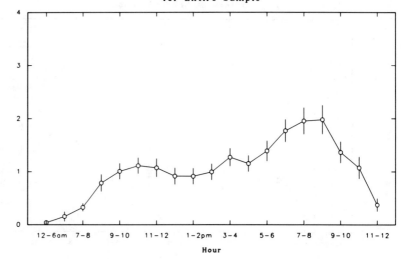

Figure 7.B.35

Mean Total Local Conversation Time per Hour
by Time of Day
Weekdays
for Entire Sample

Sample Size 573

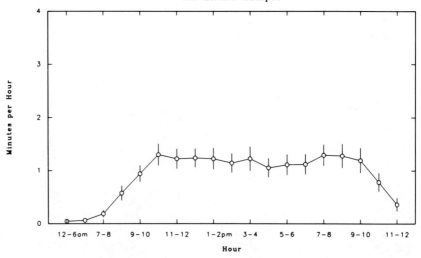

Figure 7.B.36

Mean Total Local Conversation Time per Hour
by Time of Day
Weekends
for Entire Sample

Sample Size 573

The Effects of Time of Day 205

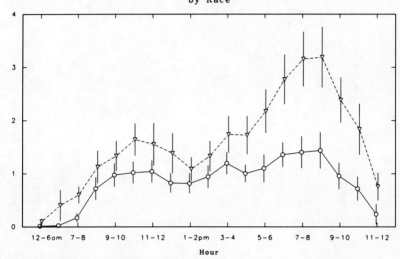

Figure 7.B.37

Mean Total Local Conversation Time per Hour
by Time of Day
Weekdays
by Race

○ White (Sample Size 232)
▽ Black (Sample Size 146)

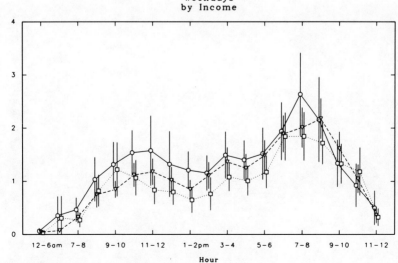

Figure 7.B.38

Mean Total Local Conversation Time per Hour
by Time of Day
Weekdays
by Income

○ Income under $5,000 (Sample Size 74)
▽ Income from $5,000 to $14,999 (Sample Size 196)
□ Income $15,000 or over (Sample Size 79)

206 The Effect of Demographics on Telephone Usage

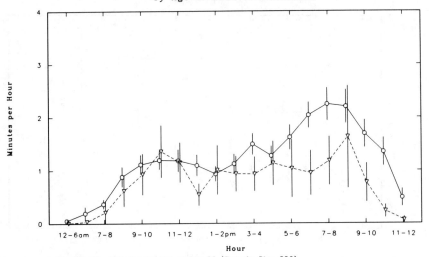

Figure 7.B.39

Mean Total Local Conversation Time per Hour
by Time of Day
Weekdays
by Age of Head of Household

○ Households with Heads Age under 65 (Sample Size 350)
▽ Households with Heads Age 65 or over (Sample Size 55)

The Effects of Time of Day 207

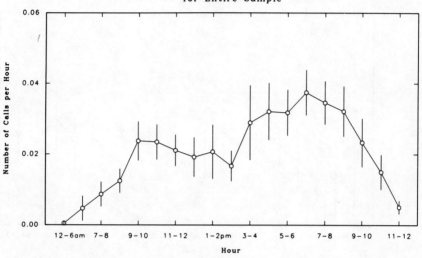

Figure 7.B.40

Mean Number of Suburban Calls per Hour
by Time of Day
Weekdays
for Entire Sample

Sample Size 573

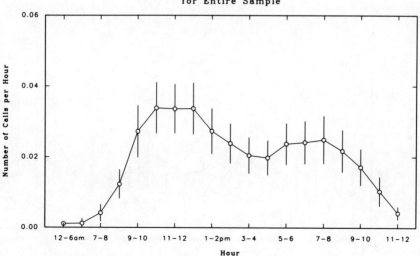

Figure 7.B.41

Mean Number of Suburban Calls per Hour
by Time of Day
Weekends
for Entire Sample

Sample Size 573

208 *The Effect of Demographics on Telephone Usage*

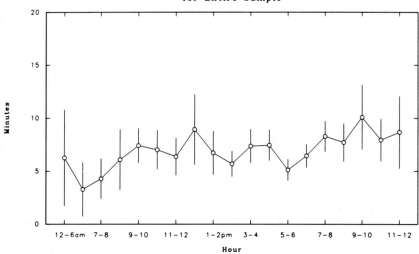

Figure 7.B.42

Mean Durations of Suburban Calls
by Time of Day
Weekdays
for Entire Sample

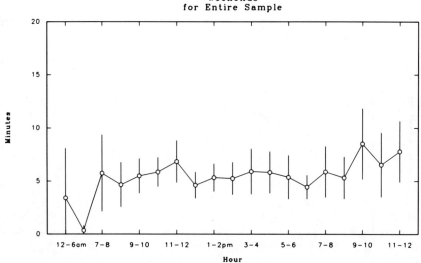

Figure 7.B.43

Mean Durations of Suburban Calls
by Time of Day
Weekends
for Entire Sample

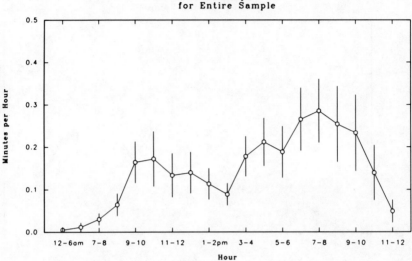

Figure 7.B.44

Mean Total Suburban Conversation Time per Hour
by Time of Day
Weekdays
for Entire Sample

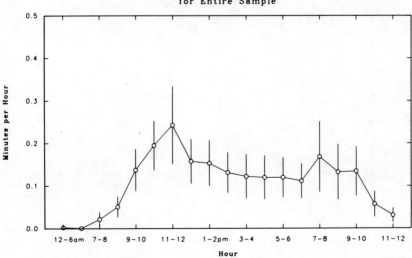

Figure 7.B.45

Mean Total Suburban Conversation Time per Hour
by Time of Day
Weekends
for Entire Sample

210 *The Effect of Demographics on Telephone Usage*

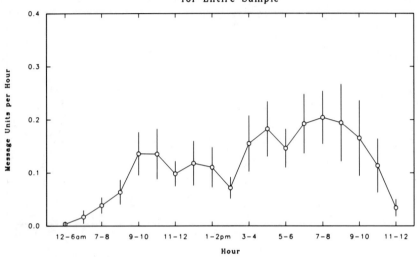

Figure 7.B.46

Mean Suburban Message Units per Hour
by Time of Day
Weekdays
for Entire Sample

Sample Size 573

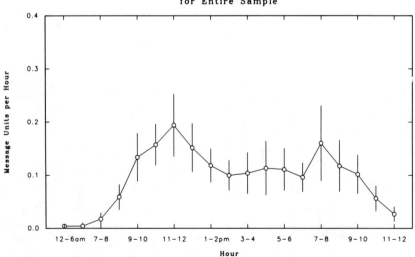

Figure 7.B.47

Mean Suburban Message Units per Hour
by Time of Day
Weekends
for Entire Sample

Sample Size 573

The Effects of Time of Day 211

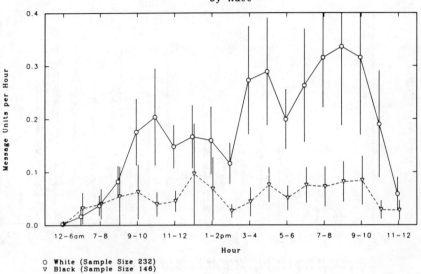

Figure 7.B.48

Mean Suburban Message Units per Hour
by Time of Day
Weekdays
by Race

○ White (Sample Size 232)
▽ Black (Sample Size 146)

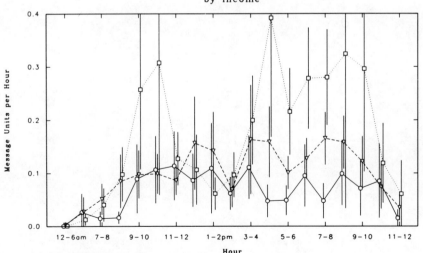

Figure 7.B.49

Mean Suburban Message Units per Hour
by Time of Day
Weekdays
by Income

○ Income under $5,000 (Sample Size 74)
▽ Income from $5,000 to $14,999 (Sample Size 196)
□ Income $15,000 or over (Sample Size 79)

212 *The Effect of Demographics on Telephone Usage*

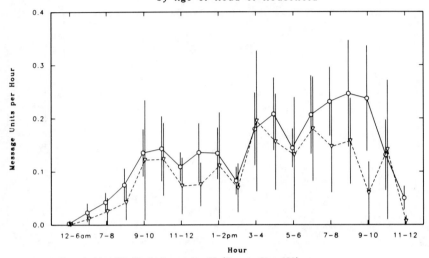

The Effects of Time of Day 213

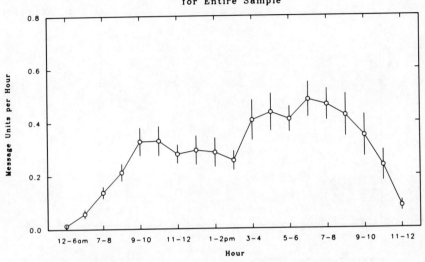

Figure 7.B.51

Mean Total Local and Suburban Message Units per Hour
by Time of Day
Weekdays
for Entire Sample

Sample Size 573

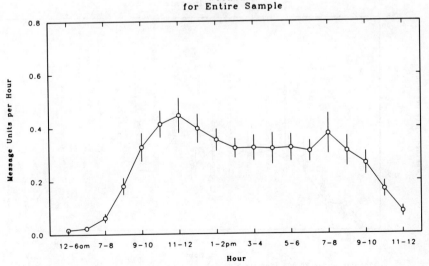

Figure 7.B.52

Mean Total Local and Suburban Message Units per Hour
by Time of Day
Weekends
for Entire Sample

Sample Size 573

214 *The Effect of Demographics on Telephone Usage*

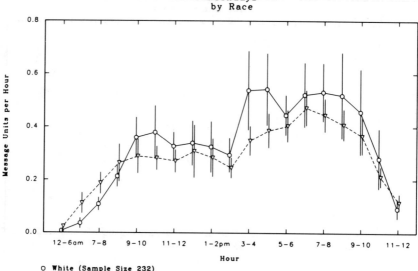

Figure 7.B.53

Mean Total Local and Suburban Message Units per Hour
by Time of Day
Weekdays
by Race

○ White (Sample Size 232)
▽ Black (Sample Size 146)

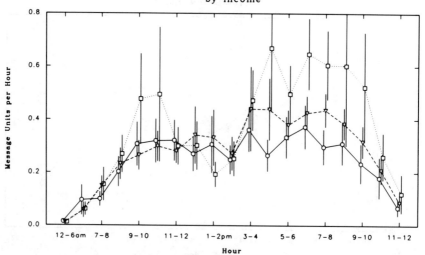

Figure 7.B.54

Mean Total Local and Suburban Message Units per Hour
by Time of Day
Weekdays
by Income

○ Income under $5,000 (Sample Size 74)
▽ Income from $5,000 to $14,999 (Sample Size 196)
□ Income $15,000 or over (Sample Size 79)

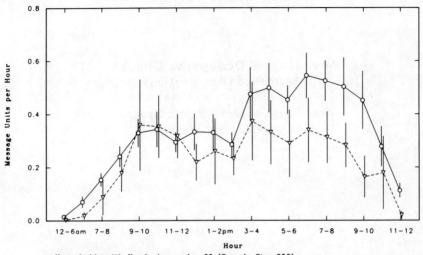

Figure 7.B.55

Mean Total Local and Suburban Message Units per Hour
by Time of Day
Weekdays
by Age of Head of Household

○ Households with Heads Age under 65 (Sample Size 350)
▽ Households with Heads Age 65 or over (Sample Size 55)

Appendix 7.C

Mean Usage of Demographic Groups Aggregated Over Weekdays and Over Weekends

Title	Page
Mean Usage of Demographic Groups Aggregated Over Weekdays	217
Mean Usage of Demographic Groups Aggregated Over Weekends	218

Mean Usage of Demographic Groups Aggregated Over Weekdays

Demographic Groups	Local			Suburban	Local and Suburban
	Number of Calls	Average Duration	Total Conv. Time	Message Units	Total Message Units
Race					
Black	0.188	7.31	1.281	0.042	0.230
	(0.020)	(0.66)	(0.159)	(0.016)	(0.029)
White	0.124	6.11	0.666	0.140	0.264
	(0.016)	(0.61)	(0.098)	(0.038)	(0.048)
Income					
Under $5,000	0.143	7.85	1.010	0.052	0.195
	(0.028)	(1.22)	(0.222)	(0.017)	(0.033)
$5,000 to $14,999	0.145	6.48	0.863	0.081	0.225
	(0.018)	(0.65)	(0.109)	(0.020)	(0.033)
$15,000 or over	0.154	5.50	0.768	0.137	0.291
	(0.023)	(0.79)	(0.121)	(0.058)	(0.066)
Age of Head of Household					
Under age 65	0.157	6.43	0.947	0.105	0.263
	(0.014)	(0.46)	(0.096)	(0.026)	(0.034)
Age 65 or over	0.104	6.60	0.612	0.076	0.180
	(0.024)	(1.33)	(0.171)	(0.031)	(0.043)
Entire Sample	0.140	6.24	0.826	0.091	0.231
	(0.010)	(0.35)	(0.069)	(0.017)	(0.023)

(1.7 standard errors of the mean are in parentheses below each usage figure.)

Normal Test for Difference in Mean Usage for Demographic Groups Aggregated Over Weekdays

White/Black	**-4.25**	**-2.27**	**-5.60**	**4.01**	1.03
Under $5,000/$5,000-14,999	-0.10	1.69	1.01	-1.84	-1.13
Under $5,000/$15,000 or over	-0.51	**2.74**	1.63	**-2.39**	**-2.22**
$5,000-14,999/$15,000 or over	-0.52	1.62	1.00	-1.55	-1.51
Under Age 65/Age 65 or over	**3.30**	-0.20	**2.91**	1.22	**2.57**

(Figures in bold type indicate significance at the 5 percent level.)

Mean Usage of Demographic Groups Aggregated Over Weekends

Demographic Groups	Local			Suburban	Local and Suburban
	Number of Calls	Average Duration	Total Conv. Time	Total Message Units	Total Message Units
Race					
Black	0.211 (0.022)	7.30 (0.75)	1.310 (0.149)	0.045 (0.022)	0.255 (0.034)
White	0.102 (0.013)	5.36 (0.60)	0.469 (0.062)	0.111 (0.028)	0.213 (0.035)
Income					
Under $5,000	0.154 (0.030)	7.22 (1.23)	1.021 (0.207)	0.039 (0.017)	0.193 (0.038)
$5,000 to $14,999	0.141 (0.018)	5.76 (0.68)	0.762 (0.108)	0.068 (0.019)	0.209 (0.027)
$15,000 or over	0.141 (0.024)	5.44 (0.87)	0.667 (0.117)	0.112 (0.042)	0.253 (0.051)
Age of Head of Household					
Under age 65	0.155 (0.014)	5.74 (0.49)	0.821 (0.082)	0.089 (0.021)	0.244 (0.027)
Age 65 or over	0.094 (0.023)	6.72 (1.18)	0.645 (0.196)	0.065 (0.028)	0.159 (0.039)
Entire Sample	0.135 (0.010)	5.97 (0.44)	0.731 (0.061)	0.077 (0.014)	0.212 (0.018)

(1.7 standard errors of the mean are in parentheses below each usage figure.)

Normal Test for Difference in Mean Usage for Demographic Groups Aggregated Over Weekends

White/Black	**-7.08**	**-3.43**	**-8.87**	**3.15**	-1.47
Under $5,000/$5,000-14,999	0.65	1.77	1.89	-1.95	-0.58
Under $5,000/$15,000 or over	0.56	**2.01**	**2.53**	**-2.76**	-1.60
$5,000-14,999/$15,000 or over	-0.03	0.49	1.01	-1.62	-1.29
Under Age 65/Age 65 or over	**3.92**	-1.30	1.40	1.16	**3.06**

(Figures in bold type indicate significance at the 5 percent level.)

CHAPTER 8

The Association of Distance with Calling Frequencies and Conversation Times

Belinda B. Brandon and Elsa M. Ancmon

8.1 Introduction

The cost of providing the capacity for a call tends to rise with the call's distance between originating and terminating Central Offices: the longer is the distance of a call, the more trunking is likely to be required, and the greater is the probability that a tandem switching machine is utilized. Thus, a rate structure that related prices to costs would take call distance into account, and the relationship between call distance and demographic characteristics is of interest.

In this chapter certain residence calling patterns for each originating Central Office Area are briefly described, and the relationship between demographic characteristics and distance called is analyzed. Two graphical display techniques are used in presenting the findings: the histogram and the line graph with error bars. Only prominent highlights of the results are discussed in the subsequent sections; readers are free to study the figures in the appendices to select the quantitative results that are of significance for their particular interests. A general pattern is that for the present sample there is at most only a mild association between distance and the studied demographic characteristics.

It must be emphasized that this chapter does not purport to study the causal effect of distance on calling, as opposed to studying the association of distance with calling. Distance certainly has a causal effect on the number of calls that a household makes to various areas. But distance is also to some extent a proxy for at least three other effects. Firstly, suburban calls are priced according to the length of haul, so calls going to a distant area may be few partly because they are expensive, and not simply because of the distance. Secondly, customers have demographic communities of interest, and demographic groups with communities of interest tend to live in close geographical proximity. Thus, a person's calls may tend to go to areas of close proximity not simply because the areas are close but also because the people that the person wants to call are nearby. Thirdly, for a given mileage band the telephone subscribers whom it is possible to call fall within an annulus of thickness equal to the width of the mileage band. Initially, the number of these subscribers tends to increase with the area of the annulus, and this area increases with the annulus' radius. However, as the radius increases, the number of subscribers that can be called under the message-rate tariff eventually decreases because more of the mileage band encompasses the toll area — which is excluded from the analysis — and Lake Michigan.

This chapter is divided into nine sections. Following this introductory section, the second section describes the data that are utilized for this chapter. The third section is a description of how distances are calculated. The next section discusses the representativeness of the sample with regard to distance. Section 8.5 displays the variation between Central Office Areas of the percentage of originating traffic that terminates in areas of different proximities. The sixth section analyzes calling characteristics by mileage bands for local and suburban calls combined and by demographic variables. The next section presents some aggregate measures of calling. Distance patterns by time of day are presented in Section 8.8. The final section is a concluding note.

8.2 The Data

This chapter utilizes some of the detailed calling data collected for the billing periods that begin in April, 1974; *viz.*, the duration, time of day, and called prefix of all out-going local calls (*i.e.*, within Chicago) and suburban calls (*i.e.*, from Chicago to the suburbs) originated by each of the customers in the Chicago sample. In this chapter, we examine distance patterns for the entire sample, and we compare these patterns between two racial groups, three income groups, and two age groups.

A map showing the location of the Central Office Areas is on the following page and is labeled Exhibit 8.1. The Central Office Areas that were sampled are shaded. The numbers below the Central Office names indicate the number of completed questionnaires in each such office. The shore of Lake Michigan runs along the right side of the page, from the eastern edge of the Rogers Park Central Office to the northeastern edge of the South Chicago Office. The eastern edges of the South Chicago and Mitchell Central Office Areas coincide with the Indiana border. The remaining boundaries of Chicago abut the area of its Illinois suburbs.

8.3 Calculation of Distance and Specification of Proximity

For the purposes of this chapter, the origin of a customer's call is defined to be the Central Office building servicing the customer's telephone. Likewise, a call's terminating point is the Central Office building servicing the terminating number. The Central Office building is generally close to the population center of the Central Office Area.

A call is discussed here in two of the ways that ratemakers have considered using for tariffs: in terms of the proximity of its originating and terminating Central Office buildings and in terms of miles between Central Office buildings. This chapter uses five categories of proximity: the originating and terminating Central Office buildings may be the "same", "contiguous local", "non-contiguous local", "contiguous suburban", and "non-contiguous suburban".

Distances between calls' originating and terminating Central Office buildings are calculated in miles and grouped into 5-mile bands. The method of determining the distance between Central Office buildings was previously developed by AT&T Long Lines. The procedure assigns to each Central Office a set of vertical and horizontal coordinates, V and H. In this particular case, the V and H coordinates of the Central Office buildings are furnished by Illinois Bell. The scaling of the coordinates is chosen so that the distance in miles between the ith and j^{th} Central Office buildings is:

$$m = (0.1[(V_i - V_j)^2 + (H_i - H_j)^2])^{1/2}.$$

The minimum distance between any two different Central Office buildings in Chicago is 0.175 miles; the maximum distance between any two Central Office buildings relevant to this study is 63 miles (between one in Chicago and one in the suburbs).

Exhibit 8.1

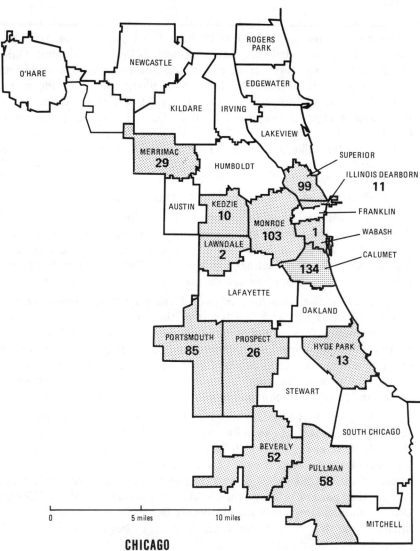

**CHICAGO
SAMPLED CENTRAL OFFICES
(WITH NUMBERS OF COMPLETED QUESTIONNAIRES)**

8.4 The Geographic Representativeness of the Sample

As discussed in Chapter 2, the sample was limited to customers served by #1 ESS or #5 Crossbar switching machines. Consequently, the demographic composition of the sampled areas differs from that of Chicago as a whole. The sampled areas also differ geographically from Chicago as a whole, as indicated by the following illustrations. Customers in Central Office Areas next to Lake Michigan had a 13 percent greater probability of being sampled than those in the rest of Chicago. Customers in Central Office Areas next to the suburbs had a 26 percent lesser probability of being sampled than the rest of Chicago. At the times when the Chicago sample was drawn (October, 1972, and April, 1973), the sample was proportional to the number of customers served by #1 ESS and #5 Crossbar machines; however, by April, 1974, this relationship no longer held. Customers in Central Office Areas adjacent to the lake were represented 42 percent more than the others, and customers in Central Office Areas bordering the suburbs were 43 percent less represented than was the rest of Chicago.

In spite of these facts, there is some evidence that the consequence of the unrepresentativeness may be minor. The following investigation of the problem was performed: Firstly, the distribution of total conversation time for the entire sample across various distances was found to be well represented by the exponential distribution (with a mean of about five miles), at least if the first two miles were aggregated. If μ is the mean distance of total conversation time, then, according to the exponential distribution, the probability density of usage for a distance d is

$$f(d) = \frac{e^{-d/\mu}}{\mu},$$

and the cumulative distribution is

$$F(d) = 1 - e^{-d/\mu}.$$

Secondly, each Central Office Area's calls were imagined to be distributed uniformly in all directions for a given distance, but with a density that declined as distance increased so as to be consistent with the exponential distribution of calls by distance as above. Using a map, the percentage of calls were tallied that the above model predicted should fall in the city, in the suburbs, "in" Lake Michigan, and in toll areas. These percentages are a set of proxies for geographical location. Thirdly, a minimum-variance estimator of the mean of customer mean distances was developed that was unbiased in these proxies. That is, a weighted mean was taken of the mean distances of the sample, where the weights were chosen so that the weighted mean of each of the proxies of the sample was equal to the mean of the respective proxy for Chicago as a whole; within all the sets of weights that satisfied that unbiasedness property, the set was chosen that minimized the variance of the estimator. It turned out that the resulting estimate differed very little from the unweighted sample mean — about one and one-half percent, or within one standard error. In addition, the weighted sample was demographically appreciably less representative of Chicago than the unweighted sample. One could also conceivably weight the sample to make it demographically representative, but there are not up-to-date data on the demographic composition of telephone subscribers in the areas of Chicago that were not sampled. Consistent with the practice for the other chapters, it was decided not to write this chapter as an attempt to provide explicit inferences for Chicago as a whole. But the authors believe that distance estimates for Chicago as a whole would differ little from those presented here.

8.5 Central Office Proximity Results

This section discusses Central Office proximity results for both local and suburban calls. Figures 8.A.1 through 8.A.4 in Appendix 8.A emphasize the variation between originating Central Office Areas in various measures of telephone usage that terminate in Central Office Areas of different proximities. Each of the four figures displays the results for all Central Office Areas together, indicated by a shaded bar on the far left, and for each of ten Central Office Areas. Read vertically, Figures 8.A.1, 8.A.3, and 8.A.4 each consist of eleven histograms, each with five bars; calls to "customer name and address",[1] "time", and "weather" information are not included. Figure 8.A.2 shows the average duration of calls for each proximity for each Area. The Hyde Park, Lawndale, and Wabash Central Offices are excluded from Figures 8.A.1 through 8.A.4 because there are no detailed calling data for any sample customers from these offices in 1974. The Central Office Areas that are not contiguous to the suburbs are indicated by a dashed line slightly below the appropriate horizontal axis.

First refer to Figure 8.A.1. For each of the five different categories of proximity, the percentage of the number of local and suburban calls terminating in Central Office Areas of that proximity is plotted on the vertical axis, and the horizontal width of the bar for each

---

[1] In Chicago, a person may call the "customer name and address" service, give a telephone number, and be told the name and address of the customer to whom that number is assigned.

Central Office Area is proportional to the square root of the total number of calls originating in that Area. (The square root of the total number of calls is used because it is roughly proportional to the standard error of the estimates as a reflection of the populations from which the sample was drawn.) One can see that 38 percent of the calls from these combined Central Office Areas terminate within the same Area. The Illinois-Dearborn and Merrimac Central Office Areas exhibit quite different calling patterns from the other Central Offices: these two offices have a considerably smaller percentage of their calling within their own boundaries, and much more to non-contiguous suburban offices. Also note that the physical proximity of sampled Central Offices to the suburbs affects the percentage of calls that go to the suburbs: as can be seen by the map in Exhibit 8.1, the Central Office Areas in Figures 8.A.1 through 8.A.4 that are contiguous to the suburbs are Portsmouth, Beverly, Merrimac, Pullman, Prospect, and Kedzie (approximately in order of decreasing contact), and these Areas tend to have larger percentages of calls terminating in the suburbs than do the others.

Average durations by Central Office Area are displayed in Figure 8.A.2. Average durations in minutes are the heights of the bars, and the square roots of the total number of calls are proportional to the widths of the bars. The average-duration results for the Illinois-Dearborn, Kedzie, and Prospect Central Office Areas are quite different from those of other Areas.

Figure 8.A.3 exhibits the results for the five categories of proximity for the percentage of total conversation time, which is plotted on the vertical axis; the widths of the bars are proportional to the square root of total conversation minutes. The Illinois-Dearborn, Merrimac, and Prospect Central Office Areas display appreciably different calling patterns from the others.

Figure 8.A.4 summarizes the findings for the percentage of total message units, which is displayed as a height. In this case the bar widths are proportional to the square root of total message units. For the percentage of total message units, Illinois-Dearborn and Merrimac differ substantially from other Central Office Areas.

8.6 Distance Results

This section of the chapter presents findings by mileage bands for local and suburban calls combined, for the entire sample and for individual demographic groups. Results are presented for the percentage of calls, average duration, percentage of total conversation time, and percentage of total message units in sub-sections 8.6.1, 8.6.2, 8.6.3, and 8.6.4, respectively. For each measure and demographic split, there are two figures, one for the cumulative distribution and one for the frequency distribution. Mileage bands are graphed on the horizontal axes and each telephone measure is on a vertical axis. For brevity, the following shorthand is used to label mileage bands: a call that is said to go zero miles is one that originates and terminates in the same Central Office Area. A call that goes "0-5 miles" is one for which the distance between originating and terminating Central Office buildings is greater than zero but less than or equal to 5 miles. The rest of the mileage bands are similarly defined, excluding the lower limit, but including the upper limit. For all but one measure of usage, the estimate is of the expected percentage of usage that falls in a given mileage band for a randomly drawn sample household. The percentage of usage falling within each mileage band is calculated for each customer in the sample, and the mean of these percentages across customers is computed for each mileage band. The exception is the case of durations, where the estimator is the sample mean of individual average durations.

In figures for the entire sample, the point estimates are bracketed by vertical bars whose length in each direction from the sample mean is 1.96 times the estimated standard error. These error bars thus provide 95 percent confidence intervals around the mean. In figures that compare groups, the size of the error bars, as in Chapter 7, are 1.7 standard errors above and below the mean. Thus in a case in which one is examining a particular, pre-chosen mileage band, one can reject the hypothesis that two independent groups from which samples were drawn have the same mean at (roughly) the 5 percent significance level if the error bars for one group do not overlap with those of the other. But the error bars are provided merely as a rough visual guide; statements about statistically significant differences are based on more precise two-tailed Normal tests with 5 percent significance levels, assuming that the sample sizes are sufficiently large to estimate the standard errors precisely.

While there are some statistically significant differences that are reported below, many of these differences are small in absolute terms. But the reader should keep two things in mind. Firstly, for the distant mileage bands, a difference between groups that is small in absolute terms can be very large in relative terms (*e.g.*, a one percentage point difference is a difference of a factor of two if one group's estimate is one percent and the other's is two percent). Secondly, differences in absolute terms between groups matter much more in terms of their cost effects for distant mileage bands than for nearby mileage bands.

One final comment may be useful preparation for the reader before he sees the distance results. Some confusion can be avoided by realizing that there is no necessary connection between the average distance of a group's calls and the group's propensity to make suburban versus local calls. This paradox is resolved by observing that a person can call a long distance within Chicago and that a person may tend more to call the suburbs if he is near them.

8.6.1 Number of Local and Suburban Calls

Figures 8.A.5A through 8.A.8B display the results for the mean percentage of total local and suburban calls by mileage bands. Figures 8.A.5A and B graph the results of total calls for the entire sample. Figure 8.A.5A shows that a mean of 36.4 percent of calls terminate within a customer's own Central Office Area. Further, 67.3 percent terminate within 5 miles; *etc.* Beyond giving the same information for zero miles, Figure 8.A.5B demonstrates that a mean of 30.9 percent of calls terminate in the 0-5-mile band; 19.8 percent in 5-10 miles; *etc.*

Figures 8.A.6A and B display results for the black and white racial groups. Black customers make a statistically significantly smaller mean percentage of calls to the 15-20-, 20-25-, 25-30-, and over-35-mile bands compared to white customers; this tendency is compensated by black customers' having a significantly larger percentage in the 0-5-mile band.

Results by income groups are portrayed in Figures 8.A.7A and B. The three income groups are well-ordered in the cases of certain mileage bands. The higher is the income, the smaller is the percentage of calls that is made to the 5-10-mile band, and the larger is the percentage of calls that is made to each of the 5-mile bands in the 10-25-mile range. The only significant difference is in the 15-20-mile band, where the under-$5,000 group is lower than the $5,000-14,999 group. These patterns are presumably caused by higher-income groups' making more calls to the suburban exchanges.

The last figures in this sub-section, 8.A.8A and B, graph the results for age of the head of household. Households headed by persons over 65 years of age are estimated to make only 31.2 percent of their calls within their own Central Office Area compared to 37.4 percent for the others, and the difference is statistically significant. The households with older heads also make a significantly larger percentage of their calls to the 5-10-mile band compared to the others.

8.6.2 Average Durations of Local and Suburban Calls

Figures 8.A.9A through 8.A.12B graph means of the sampled customers' average durations of local and suburban calls by mileage bands. As in Chapter 5, the estimated bias is removed from the durations.

In Figure 8.A.9B the means of average durations rise from 5.0 minutes for the zero-mile calls to a maximum of 7.3 minutes in the 5-10-mile band. Their lowest point — 2.6 minutes — is in the over-35-mile band.

In Figure 8.A.10B, one sees that black customers have significantly longer average durations than white customers for the 0-5- and 5-10-mile bands, peaking at 9.2 minutes for 5-10 miles. Also one sees that black customers have shorter durations than white customers for long distances. White customers' average durations peak in the 15-20-mile band at 7.7 minutes.

Results by income groups appear in Figures 8.A.11A and B. For distances up to 10 miles, the higher is the income, the lower is the mean duration. (The difference between the highest-income group and either of the other two groups is significant for the 5-10-mile band.) There is no clear pattern for greater mileages, although in the 20-25-mile band the lowest-income group has significantly shorter durations than the middle-income group.

The last figures in this sub-section — 8.A.12A and B — display the results for the age of the head of household. Households headed by persons 65 years of age or older have shorter durations for distances greater than 25 miles, and the differences are significant for the 25-30- and over-35-mile bands.

8.6.3 Total Local and Suburban Conversation Time

Results for the mean percentage of total conversation time are presented in Figures 8.A.13A through 8.A.16B. In Figure 8.A.13B a mean of 30.8 percent of total conversation time terminates in the same Central Office Area as it originates. (For the number of calls, the figure had been 36.4 percent.) As the distance from a customer's own Central Office Area increases, his total conversation time falls off less rapidly than his number of calls does.

Figures 8.A.14A and B display the results by racial groups. A larger proportion of black customers' usage is in mileage bands greater than zero and up to 10 miles than that of white customers is; the difference for the 0-5-mile band is statistically significant. Black customers even have a greater portion of their total conversation time in the

0-5-mile band than in their own Central Office Area. Black customers have a significantly smaller fraction of their conversation time terminating in the 15-20-, 25-30-, and over-35-mile bands than do white customers.

Total conversation time is presented in Figures 8.A.15A and B by income groups. Figure 8.A.15B shows no significant differences between any of the pairs of income groups.

The last figures in this section — 8.A.16A and B — study the age of the head of household. In Figure 8.A.16B, as compared to the younger group, the 65-or-over group is significantly less in its proportion of total conversation time that goes zero miles but appears higher in its proportion of usage in the 5-10-mile band (the difference is just short of being statistically significant).

8.6.4 Total Local and Suburban Message Units

This subsection discusses the mean percentage of total local and suburban message units on the basis of Figures 8.A.17A through 8.A.20B. Figure 8.A.17B demonstrates that customers make even a smaller percentage of their message units — 28.9 — in their own Central Office Area than they do for total conversation time.

Black customers, as shown in Figure 8.A.18B, have a significantly higher usage of message units in the zero- and 0-5-mile bands than do white customers, but have a significantly lower message unit usage for the 10-15-, 15-20-, 20-25-, and over-35-mile bands.

Refer to Figure 8.A.19A and B for comparisons between income groups. The only significant differences between income groups are for the 10-15-mile band, where the highest-income group has a significantly greater message unit usage than the other two groups do.

Figure 8.A.20B portrays the relationship between the age of the head of household and proportion of message units. Households with older heads have significantly lower message unit usage in their own Central Office Area than do households with younger heads. The message unit usage of the older group peaks in the 5-10-mile band.

8.7 Aggregate Measures of Local and Suburban Calling

This section of the chapter presents some comparisons between groups in their aggregate numbers of calls, average durations, total conversation times, message units, average distances, total minute-miles, and average minute-miles per call. For each measure the sample mean across customers is calculated for the same demographic groups as in the previous section. For the comparison between each pair of demographic groups, a test for a difference in means is performed. A two-tailed Normal test is used, assuming that the sample sizes are sufficiently large for the variance of each individual group to be precisely estimated. The numerical results are shown in tables in Appendix 8.B, where any significant difference is emphasized by displaying the z-statistic in bold type. The reader should keep in mind three differences between this section and Chapter 5: Firstly, the results presented in Chapter 5 for aggregate measures of calling were *medians;* the results presented in this chapter are *means*. Secondly, most of Chapter 5 separated local calls from suburban calls while this section aggregates them. Thirdly, this chapter excludes all calls for information on time, weather, and customer name and address.

Refer to Table 8.B.1 for the results on the mean of the monthly number of local and suburban calls. The value for the entire sample is 108 calls. The number made by black customers — 142 — is significantly greater than for white customers — 99 — at better than the 95 percent confidence level. There are no significant differences between the income groups; however, the higher is the income, the higher is the mean number of calls. Households headed by persons 65 years of age or older make significantly fewer calls than do those headed by younger persons — 81 versus 122 calls.

The mean of the entire sample's average durations is shown in Table 8.B.2 to be 6.5 minutes. Again, the two races are significantly different in their calling characteristics: the average duration for black customers is 7.5 minutes, and for white customers, 6.2. Durations fall monotonically as income rises, and the under-$5,000 income group has significantly longer average durations than the $15,000-or-over income group does — 7.6 versus 5.8 minutes. Households with older heads have slightly longer durations than the others do, but the difference is not significant.

In Table 8.B.3, the mean total conversation time for the entire sample is 651 minutes in a month. Total conversation time is significantly higher for black than for white customers. While it monotonically falls as income rises, the difference between no pair of income groups is significant. But total conversation time is significantly higher for younger-headed households than for older-headed households.

The results on total local and suburban message units are listed in Table 8.B.4. For the entire sample the mean is 159 message units. Interestingly, whereas Chapter 5 reported that the *median* message unit usage for black customers was estimated to be higher than for white customers, though insignificantly so, the estimate of the *mean* for black customers is lower than for white customers, though again the difference is not statistically significant. Message unit usage rises monotonically as income rises even though the number of calls seems to have the opposite pattern. The difference in usage between the lowest- and highest-income groups is statistically significant. Households headed by a senior citizen have significantly lower usage than the others do.

Table 8.B.5 reports the mean of average distances of each group's local and suburban calls. For example, for the entire sample the estimate is 4.42 miles. It is interesting that there is little variation in average distance between groups and that there are no significant differences at the 95 percent confidence level; however, the mean for black customers is significantly smaller than for white customers at the 90 percent confidence level.

The mean of the total minute-miles for a group is calculated by first determining the minute-miles for each call made by each customer in the group (*i.e.*, minutes of duration times miles), summing the minute-miles for all their calls, and then dividing by the number of customers in the group. In Table 8.B.6 one can see that the means of total minute-miles for the entire sample is 3091 in the month. Black customers have a significantly higher mean than white customers do, but there are no other significant differences.

Table 8.B.7 reports the mean of customers' average minute-miles per call, where the latter is calculated by dividing each customer's total minute-miles by his number of calls. For the entire sample, the mean of average minute-miles per call is 33.2. As for demographic groups, values of average minute-miles vary only modestly, and there are no significant differences between groups.

8.8 Distance Results by Time of Day

This section of the chapter examines distance calling patterns by time of day. Average distance, total minute-miles, and average minute-miles are examined for the entire sample for both the weekdays and weekends; the three measures are also examined for demographic groups for weekdays only. The analysis parallels that in Chapter 7 on patterns by hour of the day. As in Section 8.6, each estimated mean for any demographic group is plotted with vertical lines representing 1.7 standard errors of the mean. But for the entire sample, the size of each error bar is 1.96 standard errors. All figures for this section are in Appendix 8.C.

8.8.1 Average Distance by Time of Day

Figure 8.C.1 displays the results for the entire sample for weekdays. According to this figure, average distance is estimated to rise from early morning until about noon and then declines mildly until 6 p.m. After that, it rises again to a peak of 4.8 miles at 10-11 p.m., then falls precipitously. The average distance for most hours of the day is approximately 4 miles. The results for the entire sample for weekends are presented in Figure 8.C.2. The average distance peaks during 8-9 p.m. at 4.5 miles. As in the case of weekdays, the average distance is usually about 4 miles.

Figure 8.C.3 displays the results for the black and white races. In the evening, white customers appear to have longer average distances than black customers do. The peak for white customers is at 9-10 p.m., and for black customers is at noon-1 p.m.

Turn to Figure 8.C.4 for results for income groups. There appears to be no discernible pattern among the three income groups.

Results by age of the head of household are graphed in Figure 8.C.5. Households with older heads tend to call longer distances in the evening.

8.8.2 Total Minute-Miles by Time of Day

Figure 8.C.6 presents the results for total minute-miles for the entire sample for weekdays. Total minute-miles rise from early morning to 10-11 p.m., fall to 2-3 p.m., and then again rise to the weekday peak at 8-9 p.m. Usage then precipitously drops. This pattern is similar to that displayed for total local conversation time across hours of the weekdays in Figure 7.B.35. Figure 8.C.7 graphs the total minute-miles than the others do for the entire sample for the weekends. In that figure the peak occurs at 7-8 p.m. Again, the pattern is similar to that for total local conversation time across hours of weekends, as shown in Figure 7.B.36.

Results for the two racial groups are shown in Figure 8.C.8. Total minute-miles for black customers are higher than for white customers except between the hours of 3 and 5 p.m. In Figure 8.C.9, no prominent differences in the patterns of the three income groups can be seen. Households headed by persons aged 65 or older have lower total minute-miles than the others do during almost every hour of the day, as shown in Figure 8.C.10.

8.8.3 Average Minute-Miles by Time of Day

Average minute-miles by time of day for the entire sample are displayed in Figure 8.C.11. This measure of usage appears to peak at 8-9 p.m., although usage at 10-11 p.m. is only slightly lower. Figure 8.C.12 displays results for weekends. In this case, the peak is at 9-10 p.m.

Figure 8.C.13 graphs the results for the two racial groups. There is no consistent difference in patterns for average minute-miles between the two races except that black customers have higher usage in the early morning. Figure 8.C.14 again shows no pattern between the income groups. The last display in this section, Figure 8.C.15, presents the results by age of the head of household. Households with older heads have higher average minute-miles than younger heads do except for 8 a.m. to 1 p.m. and 11 p.m. to 12 midnight. Older heads have an especially prominent peak in average minute-miles at 8-9 p.m.

8.9 Conclusion

In this chapter, one sees the same kind of differences in the rate of calling between groups as displayed in Chapter 5. But there are only small differences between demographic groups in the way that distance is associated with calling.

The results reported in this chapter are conditional on the geographical arrangement of demographic groups in Chicago. Since the authors believe that demographic communities of interest dramatically affect calling patterns, a different geographical arrangement of groups would presumably yield different distance patterns of calling. Thus, the authors suggest that how demographic communities of interest affect calling patterns should be explicitly estimated before any attempt is made to predict distance patterns of calling in another city. This proposal is explained in more detail in Chapter 11.

Appendix 8.A

Local and Suburban Calling by Distance

Figure Number	Title	Page
8.A.1	Percentage of Local and Suburban Calls by Proximity by Originating Central Office Area	236
8.A.2	Average Durations of Local and Suburban Calls by Proximity by Originating Central Office Area	237
8.A.3	Percentage of Local and Suburban Total Conversation Time by Proximity by Originating Central Office Area	238
8.A.4	Percentage of Local and Suburban Message Units by Proximity by Originating Central Office Area	239
8.A.5A	Mean Percentage of Local and Suburban Calls, Cumulative by Distance for Entire Sample	240
8.A.5B	Mean Percentage of Local and Suburban Calls by Distance Bands for Entire Sample	240
8.A.6A	Mean Percentage of Local and Suburban Calls, Cumulative by Distance by Race	241
8.A.6B	Mean Percentage of Local and Suburban Calls by Distance Bands by Race	241
8.A.7A	Mean Percentage of Local and Suburban Calls, Cumulative by Distance by Income	242
8.A.7B	Mean Percentage of Local and Suburban Calls by Distance Bands by Income	242
8.A.8A	Mean Percentage of Local and Suburban Calls, Cumulative by Distance by Age of Head of Household	243
8.A.8B	Mean Percentage of Local and Suburban Calls by Distance Bands by Age of Head of Household	243
8.A.9A	Average Durations of Local and Suburban Calls, Cumulative by Distance for Entire Sample	244
8.A.9B	Average Durations of Local and Suburban Calls by Distance Bands for Entire Sample	244
8.A.10A	Average Durations of Local and Suburban Calls, Cumulative by Distance by Race	245
8.A.10B	Average Durations of Local and Suburban Calls by Distance Bands by Race	245
8.A.11A	Average Durations of Local and Suburban Calls, Cumulative by Distance by Income	246

8.A.11B	Average Durations of Local and Suburban Calls by Distance Bands by Income	246
8.A.12A	Average Durations of Local and Suburban Calls, Cumulative by Distance by Age of Head of Household	247
8.A.12B	Average Durations of Local and Suburban Calls by Distance Bands by Age of Head of Household	247
8.A.13A	Mean Percentage of Local and Suburban Total Conversation Time, Cumulative by Distance for Entire Sample	248
8.A.13B	Mean Percentage of Local and Suburban Total Conversation Time by Distance Bands for Entire Sample	248
8.A.14A	Mean Percentage of Local and Suburban Total Conversation Time, Cumulative by Distance by Race	249
8.A.14B	Mean Percentage of Local and Suburban Total Conversation Time by Distance Bands by Race	249
8.A.15A	Mean Percentage of Local and Suburban Total Conversation Time, Cumulative by Distance by Income	250
8.A.15B	Mean Percentage of Local and Suburban Total Conversation Time by Distance Bands by Income	250
8.A.16A	Mean Percentage of Local and Suburban Total Conversation Time, Cumulative by Distance by Age of Head of Household	251
8.A.16B	Mean Percentage of Local and Suburban Total Conversation Time by Distance Bands by Age of Head of Household	251
8.A.17A	Mean Percentage of Local and Suburban Message Units, Cumulative by Distance for Entire Sample	252
8.A.17B	Mean Percentage of Local and Suburban Message Units by Distance Bands for Entire Sample	252
8.A.18A	Mean Percentage of Local and Suburban Message Units, Cumulative by Distance by Race	253
8.A.18B	Mean Percentage of Local and Suburban Message Units by Distance Bands by Race	253
8.A.19A	Mean Percentage of Local and Suburban Message Units, Cumulative by Distance by Income	254
8.A.19B	Mean Percentage of Local and Suburban Message Units by Distance Bands by Income	254
8.A.20A	Mean Percentage of Local and Suburban Message Units, Cumulative by Distance by Age of Head of Household	255
8.A.20B	Mean Percentage of Local and Suburban Message Units by Distance Bands by Age of Head of Household	255

236 *The Effect of Demographics on Telephone Usage*

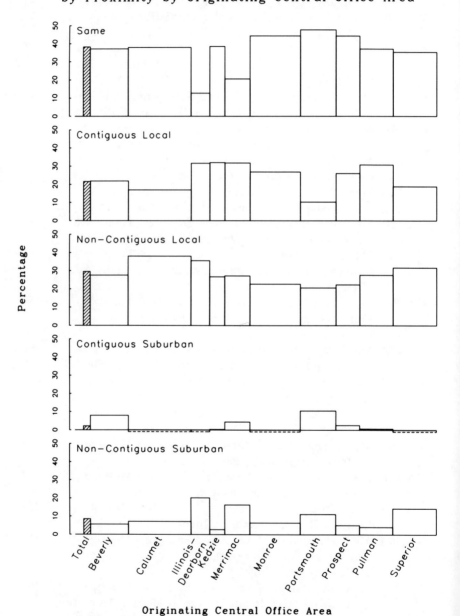

Figure 8.A.1

Percentage of Local and Suburban Calls by Proximity by Originating Central Office Area

(widths of bars are proportional to square root of number of calls)

The Association of Distance with Calling 237

Figure 8.A.2

Average Durations of Local and Suburban Calls by Proximity by Originating Central Office Area

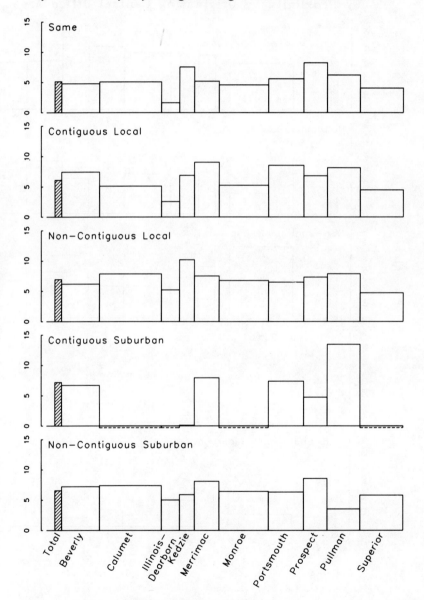

Originating Central Office Area

(widths of bars are proportional to square root of number of calls)

238 *The Effect of Demographics on Telephone Usage*

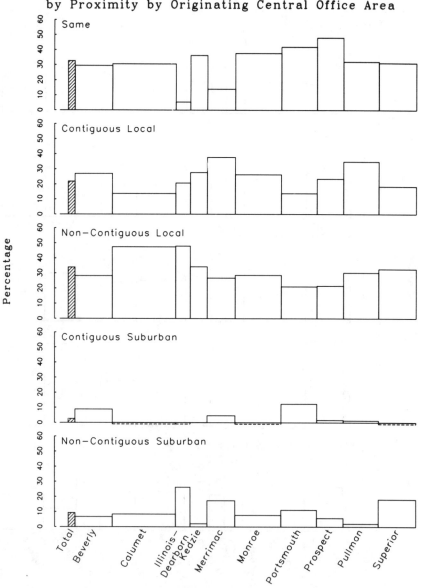

Figure 8.A.3

Percentage of Local and Suburban
Total Conversation Time
by Proximity by Originating Central Office Area

(widths of bars are proportional to square root of total conversation time)

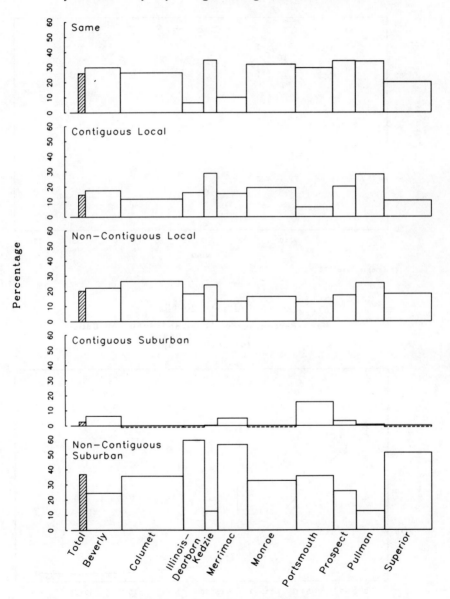

Figure 8.A.4

Percentage of Local and Suburban Message Units by Proximity by Originating Central Office Area

(widths of bars are proportional to square root of total message units)

240 *The Effect of Demographics on Telephone Usage*

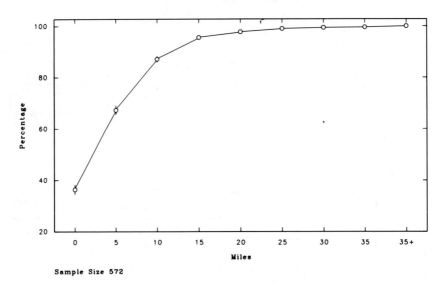

Figure 8.A.5A

Mean Percentage of Local and Suburban Calls,
Cumulative by Distance
for Entire Sample

Sample Size 572

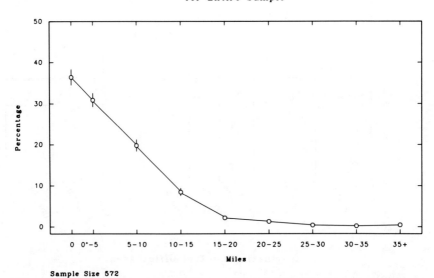

Figure 8.A.5B

Mean Percentage of Local and Suburban Calls
by Distance Bands
for Entire Sample

Sample Size 572

The Association of Distance with Calling 241

Figure 8.A.6A

**Mean Percentage of Local and Suburban Calls,
Cumulative by Distance
by Race**

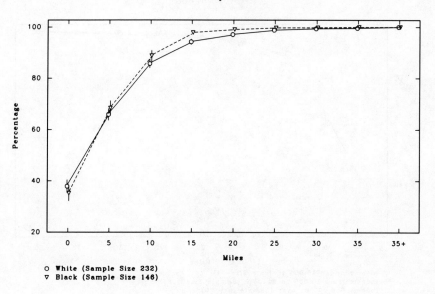

○ White (Sample Size 232)
▽ Black (Sample Size 146)

Figure 8.A.6B

**Mean Percentage of Local and Suburban Calls
by Distance Bands
by Race**

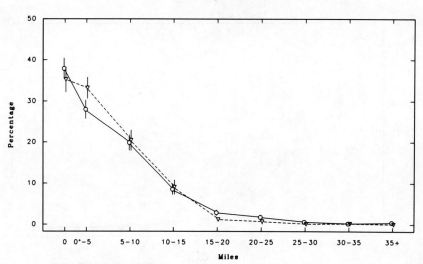

○ White (Sample Size 232)
▽ Black (Sample Size 146)

242 *The Effect of Demographics on Telephone Usage*

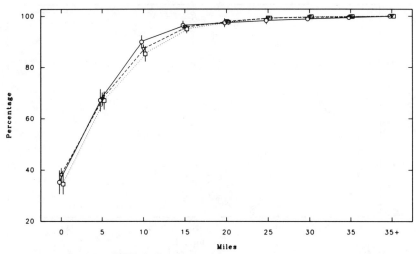

Figure 8.A.7A

Mean Percentage of Local and Suburban Calls,
Cumulative by Distance
by Income

○ Income under $5,000 (Sample Size 74)
▽ Income from $5,000 to $14,999 (Sample Size 196)
□ Income $15,000 or over (Sample Size 79)

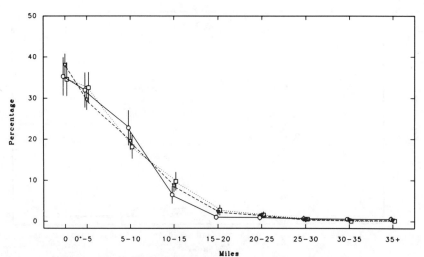

Figure 8.A.7B

Mean Percentage of Local and Suburban Calls
by Distance Bands
by Income

○ Income under $5,000 (Sample Size 74)
▽ Income from $5,000 to $14,999 (Sample Size 196)
□ Income $15,000 or over (Sample Size 79)

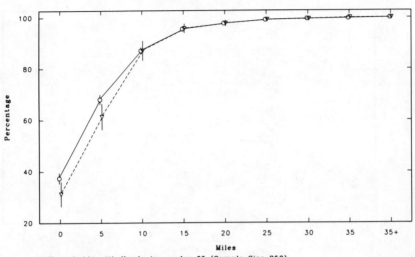

Figure 8.A.8A

Mean Percentage of Local and Suburban Calls,
Cumulative by Distance
by Age of Head of Household

○ Households with Heads Age under 65 (Sample Size 350)
▽ Households with Heads Age 65 or over (Sample Size 55)

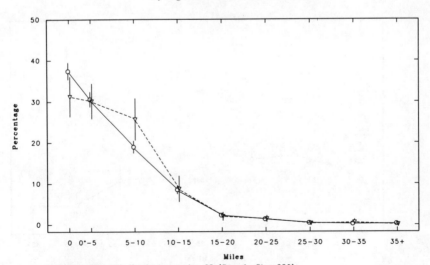

Figure 8.A.8B

Mean Percentage of Local and Suburban Calls
by Distance Bands
by Age of Head of Household

○ Households with Heads Age under 65 (Sample Size 350)
▽ Households with Heads Age 65 or over (Sample Size 55)

Figure 8.A.9A

Average Durations of Local and Suburban Calls,
Cumulative by Distance
for Entire Sample

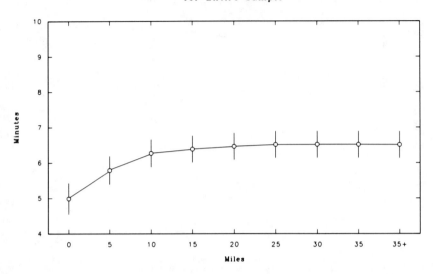

Figure 8.A.9B

Average Durations of Local and Suburban Calls
by Distance Bands
for Entire Sample

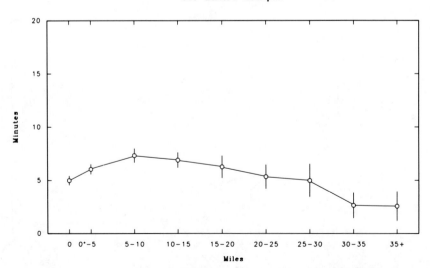

Figure 8.A.10A

**Average Durations of Local and Suburban Calls,
Cumulative by Distance
by Race**

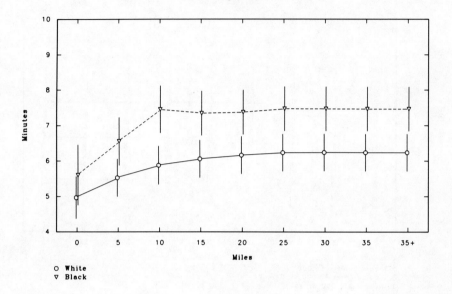

○ White
▽ Black

Figure 8.A.10B

**Average Durations of Local and Suburban Calls
by Distance Bands
by Race**

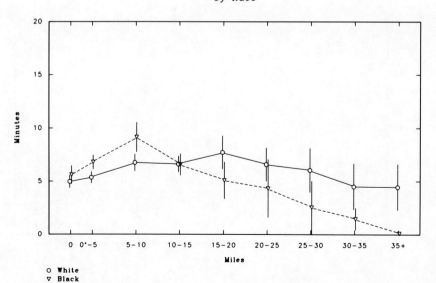

○ White
▽ Black

246 *The Effect of Demographics on Telephone Usage*

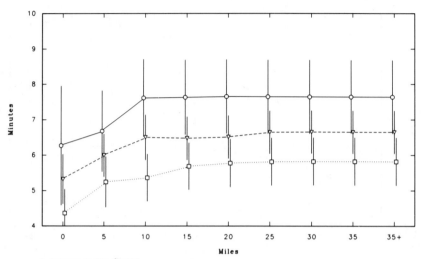

Figure 8.A.11A

Average Durations of Local and Suburban Calls,
Cumulative by Distance
by Income

○ Income under $5,000
▽ Income from $5,000 to $14,999
□ Income $15,000 or over

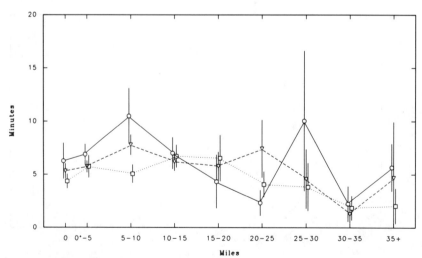

Figure 8.A.11B

Average Durations of Local and Suburban Calls
by Distance Bands
by Income

○ Income under $5,000
▽ Income from $5,000 to $14,999
□ Income $15,000 or over

Figure 8.A.12A

Average Durations of Local and Suburban Calls,
Cumulative by Distance
by Age of Head of Household

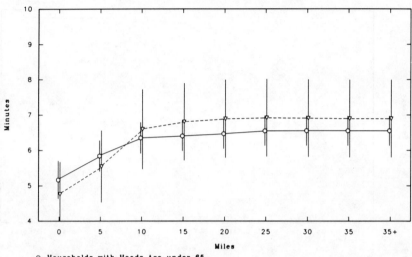

○ Households with Heads Age under 65
▽ Households with Heads Age 65 or over

Figure 8.A.12B

Average Durations of Local and Suburban Calls
by Distance Bands
by Age of Head of Household

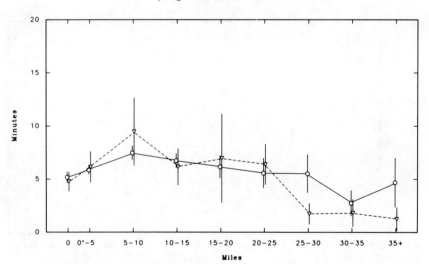

○ Households with Heads Age under 65
▽ Households with Heads Age 65 or over

248 *The Effect of Demographics on Telephone Usage*

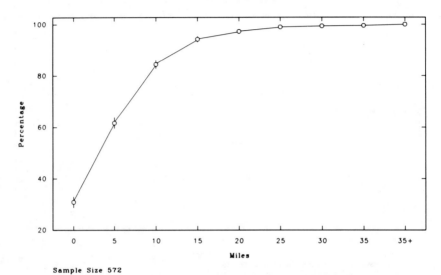

Figure 8.A.13A

Mean Percentage of Local and Suburban Total Conversation Time,
Cumulative by Distance
for Entire Sample

Sample Size 572

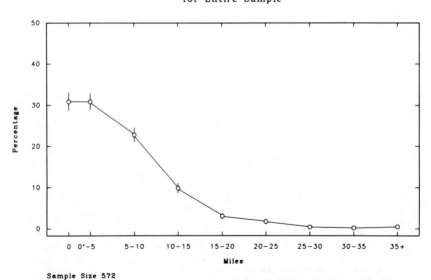

Figure 8.A.13B

Mean Percentage of Local and Suburban Total Conversation Time
by Distance Bands
for Entire Sample

Sample Size 572

Figure 8.A.14A

Mean Percentage of Local and Suburban Total Conversation Time, Cumulative by Distance by Race

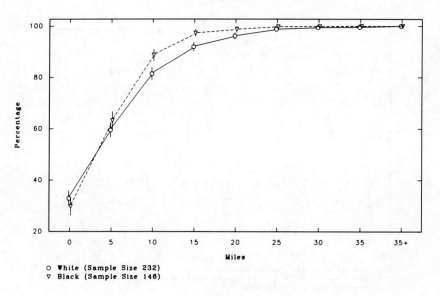

○ White (Sample Size 232)
▽ Black (Sample Size 146)

Figure 8.A.14B

Mean Percentage of Local and Suburban Total Conversation Time by Distance Bands by Race

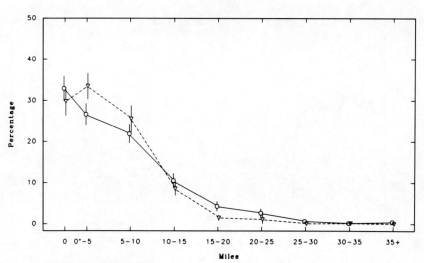

○ White (Sample Size 232)
▽ Black (Sample Size 146)

250 *The Effect of Demographics on Telephone Usage*

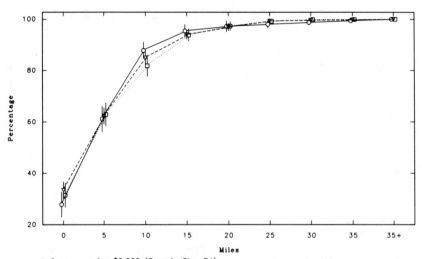

Figure 8.A.15A

Mean Percentage of Local and Suburban Total Conversation Time,
Cumulative by Distance
by Income

○ Income under $5,000 (Sample Size 74)
▽ Income from $5,000 to $14,999 (Sample Size 196)
□ Income $15,000 or over (Sample Size 79)

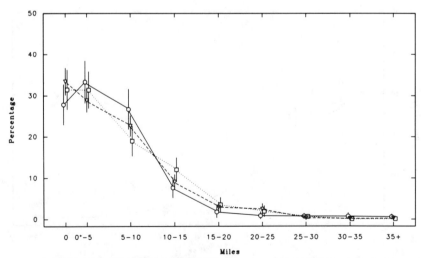

Figure 8.A.15B

Mean Percentage of Local and Suburban Total Conversation Time
by Distance Bands
by Income

○ Income under $5,000 (Sample Size 74)
▽ Income from $5,000 to $14,999 (Sample Size 196)
□ Income $15,000 or over (Sample Size 79)

Figure 8.A.16A

Mean Percentage of Local and Suburban Total Conversation Time,
Cumulative by Distance
by Age of Head of Household

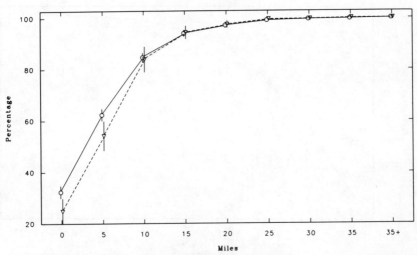

○ Households with Heads Age under 65 (Sample Size 350)
▽ Households with Heads Age 65 or over (Sample Size 55)

Figure 8.A.16B

Mean Percentage of Local and Suburban Total Conversation Time
by Distance Bands
by Age of Head of Household

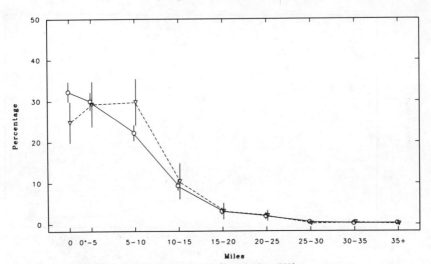

○ Households with Heads Age under 65 (Sample Size 350)
▽ Households with Heads Age 65 or over (Sample Size 55)

252 *The Effect of Demographics on Telephone Usage*

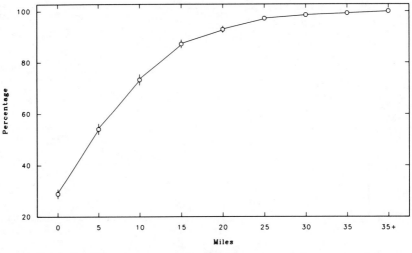

Figure 8.A.17A

Mean Percentage of Local and Suburban Message Units,
Cumulative by Distance
for Entire Sample

Sample Size 572

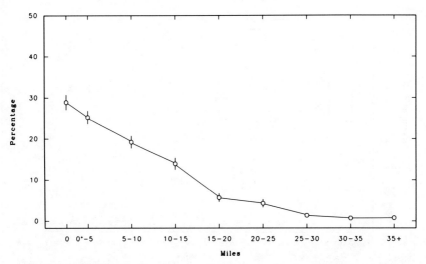

Figure 8.A.17B

Mean Percentage of Local and Suburban Message Units
by Distance Bands
for Entire Sample

Sample Size 572

Figure 8.A.18A

Mean Percentage of Local and Suburban Message Units,
Cumulative by Distance
by Race

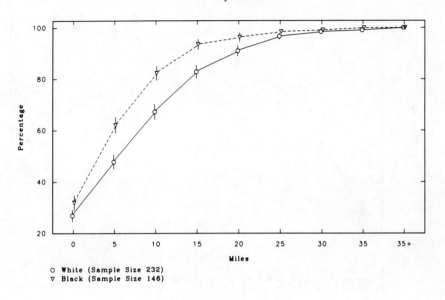

○ White (Sample Size 232)
▽ Black (Sample Size 146)

Figure 8.A.18B

Mean Percentage of Local and Suburban Message Units
by Distance Bands
by Race

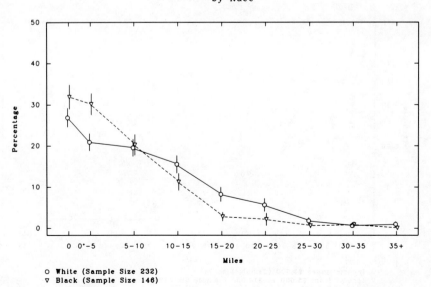

○ White (Sample Size 232)
▽ Black (Sample Size 146)

254 *The Effect of Demographics on Telephone Usage*

Figure 8.A.19A

Mean Percentage of Local and Suburban Message Units,
Cumulative by Distance
by Income

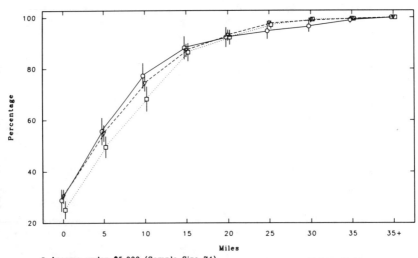

○ Income under $5,000 (Sample Size 74)
▽ Income from $5,000 to $14,999 (Sample Size 196)
□ Income $15,000 or over (Sample Size 79)

Figure 8.A.19B

Mean Percentage of Local and Suburban Message Units
by Distance Bands
by Income

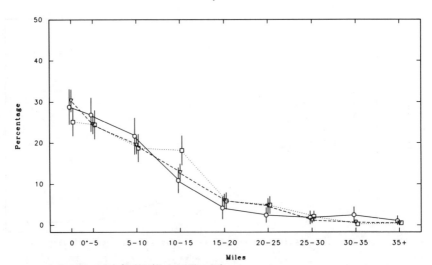

○ Income under $5,000 (Sample Size 74)
▽ Income from $5,000 to $14,999 (Sample Size 196)
□ Income $15,000 or over (Sample Size 79)

Figure 8.A.20A

Mean Percentage of Local and Suburban Message Units,
Cumulative by Distance
by Age of Head of Household

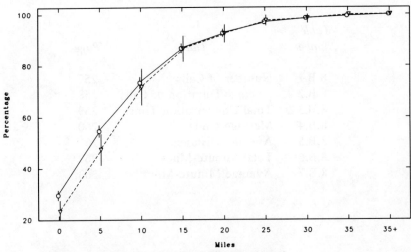

○ Households with Heads Age under 65 (Sample Size 350)
▽ Households with Heads Age 65 or over (Sample Size 55)

Figure 8.A.20B

Mean Percentage of Local and Suburban Message Units
by Distance Bands
by Age of Head of Household

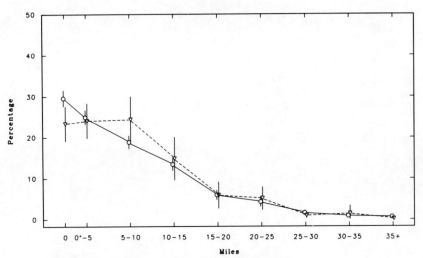

○ Households with Heads Age under 65 (Sample Size 350)
▽ Households with Heads Age 65 or over (Sample Size 55)

Appendix 8.B

Mean of Combined Local and Suburban Usage by Demographic Group

Table Number	Title	Page
8.B.1	Number of Calls	257
8.B.2	Average Duration of Calls	258
8.B.3	Total Conversation Time	259
8.B.4	Message Units	260
8.B.5	Average Distance	261
8.B.6	Total Minute-Miles	262
8.B.7	Average Minute-Miles	263

Table 8.B.1

Number of Calls

Groups	Mean	Standard Error
Race		
White	99.2	7.1
Black	142.5	8.7
Income		
Under $5,000	108.5	11.9
$5,000 to $14,999	111.8	7.7
$15,000 or over	121.0	9.9
Age of the Head of the Household		
Under age 65	122.5	6.0
Age 65 or over	80.7	10.0
Entire sample	108.0	4.4

Normal Test for Difference in Means

Groups	z-Statistic
White/Black	**-3.87**
Under $5,000/$5,000 to $14,999	-0.24
Under $5,000/$15,000 or over	-0.81
$5,000 to $14,999/$15,000 or over	-0.73
Under age 65/Age 65 or over	**3.60**

Table 8.B.2

Average Duration of Calls (Minutes)

Groups	Mean	Standard Error
Race		
White	6.23	0.31
Black	7.46	0.37
Income		
Under $5,000	7.64	0.61
$5,000 to $14,999	6.64	0.36
$15,000 or over	5.82	0.39
Age of the Head of the Household		
Under age 65	6.55	0.25
Age 65 or over	6.89	0.65
Entire sample	6.50	0.19

Normal Test for Difference in Means

Groups	z-Statistic
White/Black	**-2.56**
Under $5,000/$5,000 to $14,999	1.41
Under $5,000/$15,000 or over	**2.51**
$5,000 to $14,999/$15,000 or over	1.56
Under age 65/Age 65 or over	-0.50

Table 8.B.3

Total Conversation Time (Minutes)

Groups	Mean	Standard Error
Race		
White	548	43
Black	964	65
Income		
Under $5,000	763	89
$5,000 to $14,999	672	45
$15,000 or over	630	52
Age of the Head of Household		
Under age 65	738	40
Age 65 or over	515	73
Entire sample	651	29

Normal Test for Difference in Means

Groups	z-Statistic
White/Black	**-5.34**
Under $5,000/$5,000 to $14,999	0.91
Under $5,000/$15,000 or over	1.29
$5,000 to $14,999/$15,000 or over	0.61
Under age 65/Age 65 or over	**2.68**

Table 8.B.4

Message Units

Groups	Mean	Standard Error
Race		
White	176.2	17.9
Black	168.0	12.4
Income		
Under $5,000	137.3	13.6
$5,000 to $14,999	155.7	11.8
$15,000 or over	197.2	25.1
Age of the Head of Household		
Under age 65	181.5	12.8
Age 65 or over	123.3	16.9
Entire sample	159.1	8.6

Normal Test for Difference in Means

Groups	z-Statistic
White/Black	0.38
Under $5,000/$5,000 to $14,999	-1.02
Under $5,000/$15,000 or over	**-2.09**
$5,000 to $14,999/$15,000 or over	-1.49
Under age 65/Age 65 or over	**2.74**

Table 8.B.5

Average Distance (Miles)

Groups	Mean	Standard Error
Race		
White	4.59	0.18
Black	4.17	0.18
Income		
Under $5,000	4.40	0.32
$5,000 to $14,999	4.36	0.19
$15,000 or over	4.56	0.25
Age of the Head of Household		
Under age 65	4.38	0.14
Age 65 or over	4.78	0.31
Entire sample	4.42	0.12

Normal Test for Difference in Means

Groups	z-Statistic
White/Black	1.66
Under $5,000/$5,000 to $14,999	0.13
Under $5,000/$15,000 or over	-0.37
$5,000 to $14,999/$15,000 or over	-0.63
Under age 65/Age 65 or over	-1.19

Table 8.B.6

Total Minute-Miles

Groups	Mean	Standard Error
Race		
White	2880	382
Black	4139	315
Income		
Under $5,000	3295	376
$5,000 to $14,999	2903	229
$15,000 or over	3384	509
Age of the Head of Household		
Under age 65	3484	287
Age 65 or over	2703	384
Entire sample	3091	192

Normal Test for Difference in Means

Groups	z-Statistic
White/Black	**-2.55**
Under $5,000/$5,000 to $14,999	0.89
Under $5,000/$15,000 or over	-0.14
$5,000 to $14,999/$15,000 or over	-0.86
Under age 65/Age 65 or over	1.63

Table 8.B.7

Average Minute-Miles

Groups	Mean	Standard Error
Race		
White	33.6	2.2
Black	35.8	3.0
Income		
Under $5,000	38.5	4.1
$5,000 to $14,999	32.9	2.6
$15,000 or over	31.4	3.5
Age of the Head of Household		
Under age 65	33.2	1.8
Age 65 or over	40.7	5.2
Entire sample	33.2	1.4

Normal Test for Difference in Means

Groups	z-Statistic
White/Black	-0.60
Under $5,000/$5,000 to $14,999	1.15
Under $5,000/$15,000 or over	1.33
$5,000 to $14,999/$15,000 or over	0.36
Under age 65/Age 65 or over	-1.36

Appendix 8.C

Measure of Distance of Local and Suburban Calling by Time of Day

Figure Number	Title	Page
8.C.1	Average Distance of Local and Suburban Calls by Time of Day, Weekdays, for Entire Sample	265
8.C.2	Average Distance of Local and Suburban Calls by Time of Day, Weekends, for Entire Sample	265
8.C.3	Average Distance of Local and Suburban Calls by Time of Day, Weekdays, Race	266
8.C.4	Average Distance of Local and Suburban Calls by Time of Day, Weekdays, Income	266
8.C.5	Average Distance of Local and Suburban Calls by Time of Day, Weekdays, Age of Head of Household	267
8.C.6	Total Minute-Miles of Local and Suburban Calls by Time of Day, Weekdays, for Entire Sample	268
8.C.7	Total Minute-Miles of Local and Suburban Calls by Time of Day, Weekends, for Entire Sample	268
8.C.8	Total Minute-Miles of Local and Suburban Calls by Time of Day, Weekdays, Race	269
8.C.9	Total Minute-Miles of Local and Suburban Calls by Time of Day, Weekdays, Income	269
8.C.10	Total Minute-Miles of Local and Suburban Calls by Time of Day, Weekdays, Age of Head of Household	270
8.C.11	Average Minute-Miles of Local and Suburban Calls by Time of Day, Weekdays, for Entire Sample	271
8.C.12	Average Minute-Miles of Local and Suburban Calls by Time of Day, Weekends, for Entire Sample	271
8.C.13	Average Minute-Miles of Local and Suburban Calls by Time of Day, Weekdays, Race	272
8.C.14	Average Minute-Miles of Local and Suburban Calls by Time of Day, Weekdays, Income	272
8.C.15	Average Minute-Miles of Local and Suburban Calls by Time of Day, Weekdays, Age of Head of Household	273

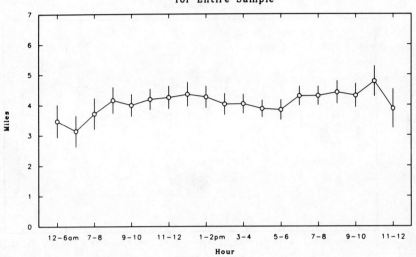

Figure 8.C.1

Average Distance of Local and Suburban Calls
by Time of Day
Weekdays
for Entire Sample

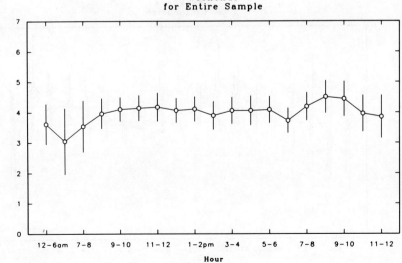

Figure 8.C.2

Average Distance of Local and Suburban Calls
by Time of Day
Weekends
for Entire Sample

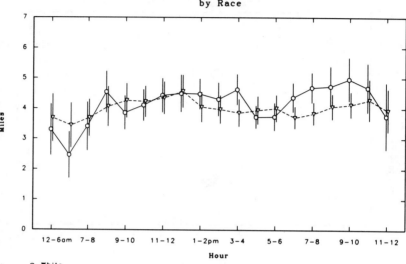

Figure 8.C.3

Average Distance of Local and Suburban Calls
by Time of Day
Weekdays
by Race

○ White
▽ Black

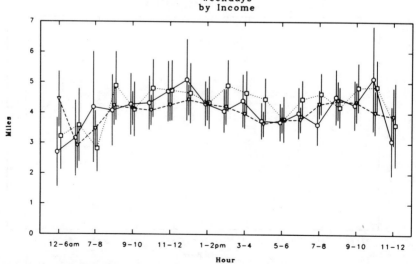

Figure 8.C.4

Average Distance of Local and Suburban Calls
by Time of Day
Weekdays
by Income

○ Income under $5,000
▽ Income from $5,000 to $14,999
□ Income $15,000 or over

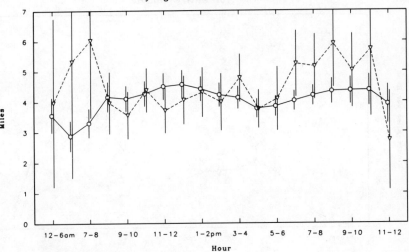

Figure 8.C.5

Average Distance of Local and Suburban Calls
by Time of Day
Weekdays
by Age of Head of Household

○ Households with Heads Age under 65
▽ Households with Heads Age 65 or over

268 *The Effect of Demographics on Telephone Usage*

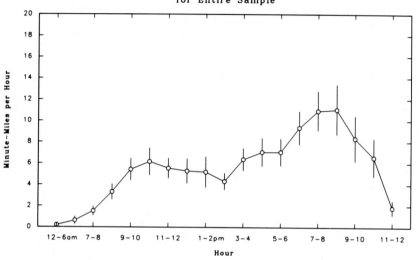

Figure 8.C.6

Total Minute-Miles of Local and Suburban Calls per Hour
by Time of Day
Weekdays
for Entire Sample

Sample Size 572

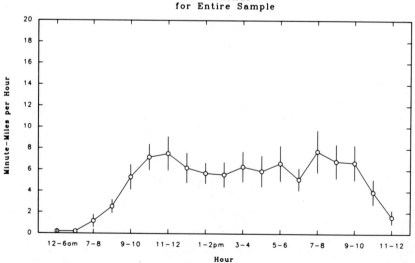

Figure 8.C.7

Total Minute-Miles of Local and Suburban Calls per Hour
by Time of Day
Weekends
for Entire Sample

Sample Size 572

The Association of Distance with Calling 269

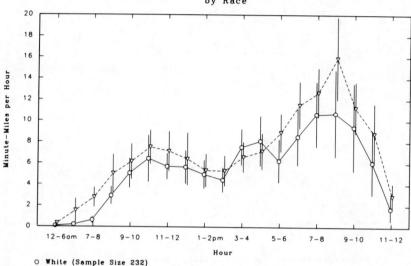

Figure 8.C.8

Total Minute−Miles of Local and Suburban Calls per Hour
by Time of Day
Weekdays
by Race

○ White (Sample Size 232)
▽ Black (Sample Size 146)

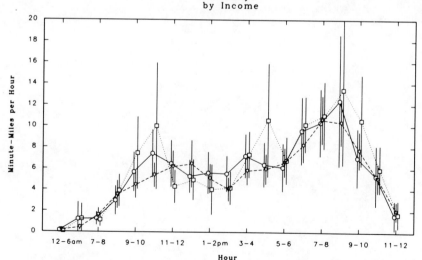

Figure 8.C.9

Total Minute−Miles of Local and Suburban Calls per Hour
by Time of Day
Weekdays
by Income

○ Income under $5,000 (Sample Size 74)
▽ Income from $5,000 to $14,999 (Sample Size 196)
□ Income $15,000 or over (Sample Size 79)

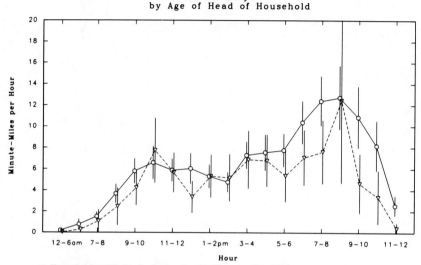

Figure 8.C.10

Total Minute-Miles of Local and Suburban Calls per Hour
by Time of Day
Weekdays
by Age of Head of Household

○ Households with Heads Age under 65 (Sample Size 350)
▽ Households with Heads Age 65 or over (Sample Size 55)

The Association of Distance with Calling 271

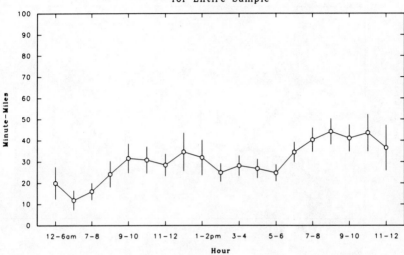

Figure 8.C.11

Average Minute-Miles of Local and Suburban Calls
by Time of Day
Weekdays
for Entire Sample

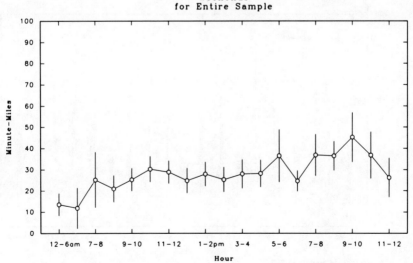

Figure 8.C.12

Average Minute-Miles of Local and Suburban Calls
by Time of Day
Weekends
for Entire Sample

272 *The Effect of Demographics on Telephone Usage*

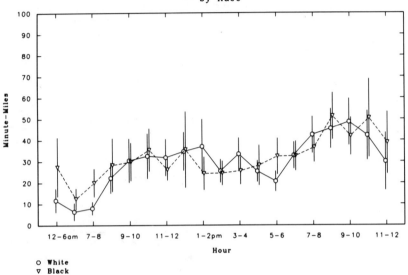

Figure 8.C.13

Average Minute-Miles of Local and Suburban Calls
by Time of Day
Weekdays
by Race

○ White
▽ Black

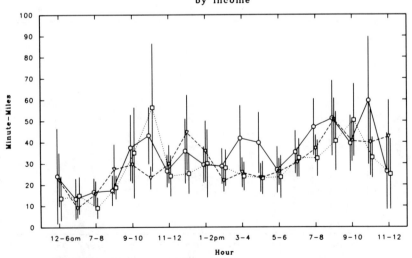

Figure 8.C.14

Average Minute-Miles of Local and Suburban Calls
by Time of Day
Weekdays
by Income

○ Income under $5,000
▽ Income from $5,000 to $14,999
□ Income $15,000 or over

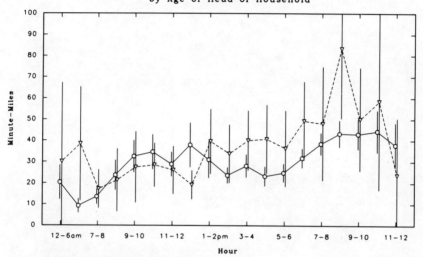

Figure 8.C.15
Average Minute-Miles of Local and Suburban Calls by Time of Day
Weekdays
by Age of Head of Household

○ Households with Heads Age under 65
▽ Households with Heads Age 65 or over

CHAPTER 9

Analysis of Billing Data

Robert H. Groff[1]

9.1 Introduction

The objective of this chapter is to examine the relationships between the customer's telephone bill and his demographic and other characteristics. We use ordinary least squares multiple regression analysis and an extended "Tobit" analysis to examine these relationships.

Analyzing telephone bills serves a useful purpose. Even if one is interested in a more detailed breakdown of telephone usage than provided by billing data (time-of-day information, for example), such detailed information is often difficult and expensive to obtain. Billing information, however, is relatively easy to obtain and can often be used as a surrogate for the more detailed information.

Following this introductory section, the second section describes the questionnaire and billing data. In the third section we discuss briefly some preliminary findings and then explain the methodology used in identifying the customers' demographic characteristics and other factors that influence the customer telephone bill. The fourth

---

[1] Several people have provided assistance with this study. I wish to thank Belinda B. Brandon and Paul S. Brandon for their guidance throughout this investigation. I also wish to thank Wayne A. Larsen, who contributed to the preliminary analyses, Edward B. Fowlkes, and William E. Taylor.

section describes the displays of results. In the concluding section of this chapter we summarize the major results of this study.

9.2 The Data

Because our main interest is in the relationship between customers' demographic characteristics and telephone bills, we use the data from the group of 623 sampled Chicago customers who answered at least one demographic question in the questionnaire described in Chapter 3.

Billing information was obtained for the 849 residence customers who were sent questionnaires. This information was gathered for the billing periods beginning in April, 1973, February, 1974, March, 1974, and April, 1974. From these data we obtained six components of the telephone bill as listed below:

(1) computed monthly bill

(2) vertical service charge

(3) basic service charge

(4) additional message unit charge

(5) toll charge

(6) unpaid balance from the previous bill.

Some of these items need further explanation. The computed monthly bill consists of the charges for telephone service during the month in question, ignoring any credit or unpaid balance that may have accrued in previous months. The vertical service component consists of charges for extras, such as Trimline® telephones, Touch-Tone® dialing, extensions, second lines, *etc.* Basic service charge refers to the charge for a telephone line, a single telephone instrument, and the customer's chosen message-unit allowance. The billing component for additional message units is the charge for those message units used in excess of the message-unit allowance. For the current analyses we look only at the computed monthly bill and the vertical service charge. Toll charges are examined in the next chapter. To date, there is no plan to study the remaining bill components.

The results that we report in detail in this chapter are for the April, 1973, data. The three months in 1974 have a substantial amount of missing data, resulting both from many customers having disconnected and from data no longer being available at the time of collection. The sample for which we have both billing and questionnaire data for April, 1973, consists of 545 customers. There are an additional nine customers for whom we have data, but their billing data are from July instead of April. These nine customers are not included in our analyses.

9.3 Methodology

The major methodology of this study was determined after many months of preliminary examinations of the data in various ways. We begin with a brief discussion of some of the preliminary results and then follow with a detailed description of the methodology used in the final models.

9.3.1 Preliminary Methodological Findings

We developed preliminary specifications using the 1973 data. We also applied those specifications to customer average bills; for each customer the average was computed over the months for which billing data were available — as many as four of them. Since for some customers only one month of billing data is available and for some as many as four months of data are available, we experimented with weighted least squares, where the weight for each customer is proportional to the square root of the number of available bills. We reached the following conclusions: (1) the indicator variable for those with only April, 1973, data did not improve the fit; (2) the difference between weighted and unweighted least squares is small; (3) customer-to-customer variability is much larger than month-to-month variability; (4) using the same set of variables that was developed for the April, 1973, data, the fits were poorer for the four months' average. This last result is not surprising. The best specification for one month may not be identically the best specification for the next month. Further, since customers' bills are strongly positively correlated across time, an averaging of each customer's bills will not, therefore, reduce the residual error appreciably.

We found that the use of dollars and cents for the computed monthly bills gives residuals from these fits that are markedly skewed, so that a few very large bills dominate the fits. By taking the square root of the reciprocal bill, we corrected this problem, finding then that the distribution of the residuals for the computed monthly bill is close

278   *The Effect of Demographics on Telephone Usage*

to a Normal distribution. For vertical service charges we do use dollars and cents. The vertical service charges, however, present a more complicated problem, as discussed in Section 9.3.3 below.

Several demographic variables, initially included in the analyses, are not contained in the final models. Both the education level and employment status of the head of household have minimal effects on the bills. After some preliminary study, billing period variables were also deleted from the final models. Since the ten billing periods are spread throughout the month, the finding of a billing period effect would not have been surprising. Instead, however, we found no significant effect attributable to the billing periods.

Several two-way interactions were also tested for significance. The hope was that the relationship between the dependent bill variables and the independent demographic variables would not be in the form of interactions. Only one interaction was significant at the 5 percent level for the transformed computed monthly bill, and another was significant for vertical service charges.

The only non-demographic characteristic we studied, other than billing period, was the Central Office Area. Such a variable may serve as a proxy for unmeasured demographic, cultural, or environmental effects. We found, however, that significant differences do not exist between all of the Central Office Areas. We were able to collapse the original 12 Central Office Areas for which we have data into three distinct Central Office groupings.

9.3.2 Computed Monthly Bill

In this subsection, we outline the methodology used in isolating customers' demographic characteristics and other effects which influence a customer's computed monthly telephone bill. Section 9.4 contains the results.

The chosen methodology for analyzing computed monthly bill is multivariate ordinary least-squares regression. We fit the following model:

$$-(y_i)^{-\frac{1}{2}} = x_i \beta + \epsilon_i, \quad i = 1,\ldots,n, \tag{9.1}$$

where

i = customer index
y_i = computed monthly bill for customer i
x_i = $(1 \times p)$ vector of independent variables for customer i
β = $(p \times 1)$ vector of parameters to be estimated
ϵ_i = independent and identically distributed error term.

The dependent variable is a power transform of the computed monthly bill in order to make the residuals close to Normally distributed. The Q-Q plot of the residuals versus their theoretical standard Normal quantiles is shown as Figure 9.A.1 in the Appendix.[2] If they were Normally distributed, then each point would lie along the straight diagonal line. (The line's slope is the estimated standard deviation of the residuals, and its intercept is at zero, their mean.) It appears that the residuals are reasonably close to Normality. The main exception is the lower tail of the distribution. If we examine the plot of the residuals versus the fitted values in Figure 9.A.2, we can see that the departure from Normality is due to the diagonal striations in the lower part of the plot. This phenomenon is explained in the following subsection on vertical service charges, where the model makes explicit recognition of such striations. For the computed monthly bill, however, we do not consider the problem to be large enough to affect the results appreciably.

All of the data on customer characteristics are discrete, or categorical. Included in the model are six demographic characteristics and a Central Office characteristic. These characteristics and their categories are listed in Exhibit 9.1. For each customer characteristic with c categories, we construct $c-1$ dummy variables which we use for the x_i data vectors in Eqn. (9.1), where the first listed category of a characteristic is implied for a customer if none of the dummy variables for that characteristic is equal to one for him.[3]

---

[2] Q-Q plots are explained in Martin B. Wilk and Ramanathan Gnanadesikan, "Probability Plotting Methods for the Analysis of Data", *Biometrika,* v. 55, no. 1 (March, 1968), pp. 1-17.

[3] In order to reduce computational rounding errors, the Helmert orthogonal transformation was applied to each set of variables. (See, *e.g.,* D. A. S. Fraser, *Statistics: An Introduction* (New York: John Wiley and Sons, Inc., 1960), pp. 173-184.) The regression was run with the transformed data. The inverse of that transformation was then applied to calculate the coefficients and their standard errors for the dummy variables which are reported herein.

Exhibit 9.1

Customer Characteristics

Household Income ($)	Age of Head of Household (Years)
under 5 thousand	under 25
5-9 thousand	25-34
9-15 thousand	35-44
15-20 thousand	45-54
over 20 thousand	55-64
no answer	65 or older

Number of Household Members at Least 10 Years of Age	Marital Status
1	single
2	married
3	separated
4	divorced
5	widowed
over 5	

Racial Group	Years Lived in Chicago
white	1-15
black	over 15
other	

Central Office Groups

1. Calumet, Merrimac, Monroe, Portsmouth, Prospect
2. Hyde Park, Illinois-Dearborn, Superior
3. Beverly, Kedzie, Lawndale, Pullman

An interaction between black households and the number of household members at least 10 years of age also enters the model. For non-black households, this interaction variable takes on an entry of zero; for black households, it takes on a value equal to the number of household members at least 10 years of age, with a maximum value of six for households with six or more such members.

9.3.3 Vertical Service Charges

The set of customer characteristics which we relate to vertical service charges is the same as for the computed monthly bill, with one exception: rather than a race-size interaction, we use an interaction between the characteristics "customers who have lived in Chicago over 15 years" and "age of the head of household". For customers who have lived in Chicago over 15 years, this interaction variable contains an entry of zero; for customers who have lived in Chicago 1 to 15 years, this interaction variable contains a value ranging from zero to five, with zero corresponding to the youngest age group (under 25) and five corresponding to the oldest age group (65 or older).

We assume that vertical service charges are linearly related to each of the independent variables:

$$y_i = x_i \beta + \epsilon_i, \quad i = 1,...,n, \tag{9.2}$$

where

- i = customer index
- y_i = the vertical service charge for customer i
- x_i = $(1 \times p)$ vector of independent variables for customer i
- β = $(p \times 1)$ vector of parameters to be estimated
- ϵ_i = independently distributed random error term.

The estimation methodology differs from that used for the computed monthly bill. Let us motivate the methodology by reference to Figure 9.A.3, which is a plot of the estimated residuals $y_i - x_i\hat{\beta}$ versus the fitted values $x_i\hat{\beta}$, where $\hat{\beta}$ is estimated with the methodology explained below. Note that in the figure there are three major striations running from the left to the lower right. The three striations consist of the large fraction of sample customers who have vertical service charges of $0, $.50, and $.75 (29 percent, 10 percent, and 13 percent of the sample, respectively). For any customer with, say, charges of $0, we know that

$$\$0 = x_i \beta + \epsilon_i,$$

so

$$\epsilon_i = - x_i \beta.$$

Thus, we know that the points in Figure 9.A.3 which correspond to the set of customers with zero charges, having a wide range of $x_i\beta$'s, will lie along a straight line with slope of -1 and intercept of $0. A similar explanation holds for the $.50 and $.75 striations. There are other striations, but we believe that they are considerably less important because the number of customers in each of the other striations is far fewer than in the major ones.

The prominent striations imply that the use of ordinary least squares regression is inappropriate in the case of vertical service charges for the present sample. The striations and the lack of points below them cause an appreciable correlation between the independent variables and the error term in Eqn. (9.2). Ordinary least squares produces biased estimated coefficients in the presence of such a correlation. "Tobit" analysis is a maximum likelihood procedure which was designed to deal with the case where an appreciable fraction of the observations is at some lower limit, and the rest of the observations are smoothly distributed above the limit.[4] Here we extend Tobit analysis by modeling *three* probability masses — at $0, $.50, and $.75. The remainder of this subsection presents this extension, which a nontechnical reader may want to skip.

Imagine that there is some unobservable, or "latent", variable y^* which is linearly related to the independent variables, and which, for given values of the independent variables, is Normally distributed; i.e.,

$$y_i^* = x_i \beta + \epsilon_i^*, \quad i = 1,...,n, \tag{9.3}$$

where ϵ_i^* is the latent error term. We assume that any customer whose y_i^* falls below zero has vertical service charges of $0; so, given $x_i\beta$, the probability that a customer has vertical service charges of zero is

$$pr(\$0) = \Phi\left(\frac{-x_i\beta}{\sigma}\right), \tag{9.4}$$

where Φ represents the cumulative standard Normal distribution function, and σ is the standard deviation of ϵ_i^* in Eqn. (9.3). So far the model coincides with the standard Tobit model. The extension is achieved by assuming that, given $x_i\beta$, the probability that a customer has a vertical service charge of $.50 is

$$pr(\$0.50) = \Phi\left(\frac{\$0.50 - x_i\beta}{\sigma}\right) - \Phi\left(\frac{-x_i\beta}{\sigma}\right), \tag{9.5}$$

and the probability of $.75 is

---

[4] James Tobin, "Estimation of Relationships for Limited Dependent Variables", *Econometrica*, v. 26, no. 1 (January, 1958), pp. 24-36.

$$pr(\$0.75) = \Phi\left[\frac{\$0.75 - x_i\beta}{\sigma}\right] - \Phi\left[\frac{\$0.50 - x_i\beta}{\sigma}\right]. \quad (9.6)$$

On the other hand, if a customer's y_i^* exceeds $.75, then he has vertical service charges equal to y_i^*. Succinctly, letting y_i represent observed vertical service charges, we have

$$y_i = \begin{cases} \$0.00 & \text{if } y_i^* \leq \$0.00 \\ \$0.50 & \text{if } \$0.00 < y_i^* \leq \$0.50 \\ \$0.75 & \text{if } \$0.50 < y_i^* \leq \$0.75 \\ y_i^* & \text{if } \$0.75 < y_i^*. \end{cases} \quad (9.7)$$

These ideas are illustrated in Exhibit 9.2 on the following page. The top graph is the probability density function of y_i^*, and the bottom one is the cumulative distribution function. The theoretical probability of vertical service charges of $0 is the area under the dotted density function from the far left to the vertical line at zero; it is also the level of the dotted cumulative distribution function at zero. The probability of $.50 is the area under the density function between the vertical lines at $0 and at $.50; it is also the difference between the cumulative probabilities at $.50 and $0. The probability of $.75 is analogously determined. The solid lines on each graph thus represent the theoretical density and cumulative distribution functions of the vertical service charges, respectively, where the heavy vertical lines on the density plot symbolize the probability masses at $0, $.50, and $.75.

Given the parameters β and σ, the likelihood of observing the data in our sample is

$$L(\beta, \sigma) = \prod_{S_0} \left[\Phi\left(\frac{-x_i\beta}{\sigma}\right)\right] \prod_{S_{50}} \left[\Phi\left(\frac{\$0.50 - x_i\beta}{\sigma}\right) - \Phi\left(\frac{-x_i\beta}{\sigma}\right)\right] \quad (9.8)$$

$$\cdot \prod_{S_{75}} \left[\Phi\left(\frac{\$0.75 - x_i\beta}{\sigma}\right) - \Phi\left(\frac{\$0.50 - x_i\beta}{\sigma}\right)\right] \prod_{S} \phi\left(\frac{y_i - x_i\beta}{\sigma}\right),$$

where S_0 is the set of observations with zero vertical service charges, S_{50} is the set with $.50, S_{75} is the set with $.75, S is the remaining set, with charges greater than $.75, and $\phi(\cdot)$ is the standard Normal probability density function. We obtain estimates of β and σ by finding their values which maximize the value of the above likelihood function.

Exhibit 9.2

Distribution of Latent and Actual Vertical Service Charges

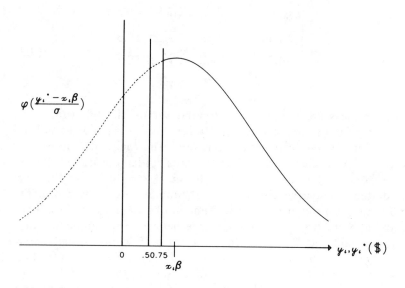

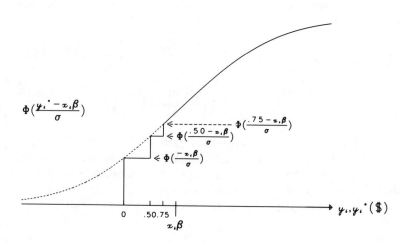

We can use the resulting estimates, if from a large sample, to obtain an unbiased prediction of vertical service charges for a household where we have data on its demographic characteristics. Letting $\hat{\beta}$ represent the estimated coefficients, the prediction is not simply $x_i\beta$, as it would be in the case of ordinary least squares.

Instead, the expected value of y_i given $x_i\beta$ is[5]

$$E(y_i|x_i\beta) = \left[1 - \Phi\left(\frac{\$0.75 - x_i\beta}{\sigma}\right)\right]\left[x_i\beta + \frac{\sigma\phi\left(\frac{\$0.75 - x_i\beta}{\sigma}\right)}{1 - \Phi\left(\frac{\$0.75 - x_i\beta}{\sigma}\right)}\right] \quad (9.9)$$

$$+ \left[\Phi\left(\frac{\$0.75 - x_i\beta}{\sigma}\right) - \Phi\left(\frac{\$0.50 - x_i\beta}{\sigma}\right)\right]\$0.75$$

$$+ \left[\Phi\left(\frac{\$0.50 - x_i\beta}{\sigma}\right) - \Phi\left(\frac{-x_i\beta}{\sigma}\right)\right]\$0.50 \ .$$

The expected value of vertical service charges exceeds $x_i\beta$, but the difference is much larger for small values of $x_i\beta$ than for large values. As $x_i\beta$ rises, the expected value rises, but not so rapidly. Figure 9.A.4 in the Appendix displays the functional relationship of the expected value of vertical services with $x_i\beta$, using the value of σ estimated for the model. The straight diagonal line is for reference, showing where the function would be if each expected value were to equal its respective fitted value. Here are some illustrative points on the function:

---

[5] From Norman L. Johnson and Samuel Kotz, *Distributions in Statistics: Continuous Univariate Distributions — 1* (Boston: Houghton Mifflin, 1970), p. 81, we know that the mean of a Normal distribution truncated at $.75 is

$$x_i\beta + \frac{\sigma\phi\left(\frac{\$0.75 - x_i\beta}{\sigma}\right)}{1 - \Phi\left(\frac{\$0.75 - x_i\beta}{\sigma}\right)} \ .$$

Eqn. (9.9) is obtained by multiplying the above expression by the probability that y_i is above $.75, plus $.75 times its probability, plus similar terms for $.50 and $0.

$x_i\beta$	$E(y_i\|x_i\beta)$	$E(y_i\|x_i\beta) - x_i\beta$
-$2.00	$0.18	$2.18
-1.00	0.43	1.43
0.00	0.83	0.83
1.00	1.43	0.43
2.00	2.19	0.19
3.00	3.07	0.07

9.4 Regression Results

We now discuss the results of our analyses, which are based on the regression procedures of Sections 9.3.2 and 9.3.3. We begin with a look at the results for the computed monthly bill and follow with a discussion of the results for vertical service charge.

9.4.1 Results for Computed Monthly Bill

The results for the computed monthly bill are presented in Figure 9.B.1 in the Appendix. We see that only a small amount of the total variation between individual households is explained by the demographic variables; the R^2 is 0.240. We lack, therefore, good predictive power for individual households; but, as discussed in Chapter 6, we may still be able to predict aggregates with good accuracy. The entire model is highly significant, with an F-statistic of 6.55 (the critical value of the F at the one percent significance level is less than 1.9 when there are 25 and 519 degrees of freedom). The set of demographic variables also adds significant explanatory power to a regression whose only independent variables are a constant term and the variables for the Central Office groups. The remainder of the table shows all the customer characteristic subgroups and their sample sizes, estimated coefficients, their standard errors, and their t-values.

All the numerical results in the table pertain to the regression where the negative of the square root of the reciprocal telephone bill is the dependent variable. Although we have a linear model for the transformed bill data, the corresponding model for the untransformed bill (in dollars) is non-linear with interactions among all the variables. Because of the computational difficulty, no statistical tests concerning the bill, in dollars, are presented here. However, as an example of the magnitude of the differences in bills represented by these coefficients, below we give approximate predicted computed monthly bills, in dollars, for each of the income subgroups, for a set of households which is imagined to be average in all respects other than income.

Income	Predicted Bill
Under 5 thousand	$17.87
5-9 thousand	$19.44
9-15 thousand	$19.71
15-20 thousand	$21.90
Over 20 thousand	$26.89
No answer	$21.78

The approximate predicted monthly bill for customers who are average in all respects is $20.99. For each of the above predictions we removed most of the bias which results from the model's expressing bills as a non-linear function of random variables.[6] The reader can get a rough idea of the dollar differences which are implied by other coefficients in the table by multiplying each coefficient by 200.

Now consider the results for each customer characteristic. We use the 5 percent level as our standard of statistical significance. For the income variables, a one-tailed Normal test is used, because of the hypothesis that bills are expected to rise with income. For all other variables, a two-tailed test is used. In the table, those coefficients that are significant are shown in boldface type. The main results are these:

(1) We find that the computed monthly bill estimated by the model is a monotonically increasing function of income. The significant differences are between the over-$20,000 group and the groups under $15,000, as well as between the $15,000-to-$20,000 group and the under-$5,000 group.

(2) The "other" racial group is little different from the white group, other things equal. But the black group has significantly larger bills than either the white or the "other" group (the test for the latter is not displayed). At this point, however, we must take note of the

---

[6] From Eqn. (9.1),
$$y_i = [(x_i\beta + u_i)^{-2}].$$

Taking the expected value of a 5-term Taylor approximation around $x_i\beta$ of the right-hand side of the above equation yields

$$E(y_i) \approx (x_i\beta)^{-2} + 3\ \sigma^2(x_i\beta)^{-4} + 5\ \sigma_4(x_i\beta)^{-6},$$

where σ_4 is the fourth moment of the Normal distribution. Neglecting the variance of $\hat{\beta}$, which is small relative to σ^2, we obtain predicted bills by substituting estimated values for β, σ^2, and σ_4 in the above equation.

significant negative coefficient on the variable which is listed at the bottom of the second page of the table — the interaction of black households and household size. Its negative sign implies that the difference between the black and white groups is smaller for larger households. In fact, it can be shown that the estimate for this simple interaction implies that the difference between the races is insignificant for households with five or more members over 10 years old.

(3) Households whose heads have lived in Chicago over 15 years have significantly lower bills than do those who have lived there a shorter time.

(4) Because the regression contains the interaction of black households and household size, coefficients for the number of household members at least 10 years of age show the effect of household size for the white and "other" racial groups. We see that for them the monthly bill rises monotonically as household size rises. The estimated effect of this customer characteristic is larger than that of any other. The statistically significant differences are between either the one- or two-member households on the one hand and any of the larger sizes on the other. As for black households, the coefficients for household size and for the interaction combined imply that bills are significantly higher for three-member households than for either one- or two-member households.

(5) The bill decreases monotonically as the age of the head of household increases. The 55-to-64-year-old group has a significantly lower estimated bill than the groups of age 34 or less. Also the oldest group, 65 years or older, has a significantly lower bill than each group of age 54 or less.

(6) We find bills of two of the Central Office groupings to differ significantly from one another. This shows that these Central Office groupings have important qualities, other than reflected by the household characteristics that are already in the model. The groupings may serve as proxies for important left-out variables.

9.4.2 Results for Vertical Service Charge

As explained in Section 9.3.3 above, we applied an extended Tobit model to the data on vertical service charges ("VSC's"). The maximum likelihood estimates of the parameters of the model are presented in Figure 9.B.2. Also shown are the traditional multiple correlation coefficient (R^2) and residual standard errors; they are computed by taking each residual as

$$res_i = y_i - E(y_i | x_i \hat{\beta}),$$

where the right-most term is calculated from Eqn. (9.9). As in the case of the computed monthly bill, the ability of the model to explain the variation between individual households in their VSC's is low, and the residual standard error is high. But the model can still probably predict the VSC's of large groups with good accuracy. There are large differences between groups, and the hypothesis that none of the demographics contribute anything compared to a naive model can be rejected by a likelihood ratio test at better than the 99.5 percent confidence level ($\chi^2 = 82$, compared with a critical value of less than 45 for 23 degrees of freedom).[7]

When studying the results, the reader should remember that the estimated coefficients do not directly give the precise dollar difference in VSC's between groups, although they do give the correct sign of the difference. For example, suppose that some household of given characteristics, including an income of under $5,000, has a fitted value of $x_i\beta$. Also suppose that another household has identical characteristics except for having an income of over $20,000. The coefficient for that income subgroup implies that the second household has a fitted value of $x_i\beta + 1.007$. The implication of Section 9.3.3 is that the expected VSC for the second household is higher than that of the first, but not as much as $1.00 higher. Figure 9.A.4 gives the precise relationship between the expected value and the fitted value. In the above example, if $x_i\beta$ for the first household is -$2.00, then the second household's expected VSC is only $.25 higher than the first's. The following table shows that instance, together with a few other cases:

Fitted Value ($)		Difference in Expected
No. 1	No. 2	
-2.00	-1.00	0.25
-1.00	0.00	0.40
0.00	1.00	0.60
1.00	2.00	0.76
2.00	3.00	0.87

These figures provide a rough guide for adjusting the coefficients so as to tell what dollar differences the coefficients imply. It may be useful to know that the mean fitted value is a little over $1.00.

---

[7] The independent variables in the naive model are a constant term and the two variables for Central Office Area groups.

There are a large number of statistically significant differences between groups. Some of the differences are also large in relative terms — even the same magnitude as the VSC sample mean of $1.48 per month. In detail, the results are these:

(1) The VSC tends to rise as income rises. Again using the one-tailed Normal test for the income variables, all income groups over $5,000 have a significantly higher VSC than the under-$5,000 group.

(2) Black customers have a significantly higher VSC than white customers, other things equal.

(3) Observing the effect of years lived in Chicago is complicated by the interaction between that characteristic and the age of the head of household. For the households with heads under 45 years of age, having lived in Chicago over 15 years increases the VSC.

(4) We estimate that the VSC rises as the number of members of the household at least 10 years of age increases up to three members, then declines as the number of members increases further.

(5) For households with heads who have lived in Chicago over 15 years, the coefficients for the age of the head of household imply that the VSC rises as age increases to 35 to 44 years, and then decreases for the older age groups. While no group is significantly different from the under-25 group, the 25-to-34-year-old group has a significantly higher VSC than the oldest three groups, and the 35-to-44 group has a significantly higher VSC than the 65-or-older group. The age coefficients together with the coefficient on the interaction variable imply that, for households with heads who have lived in Chicago 1 to 15 years, the under-25-year-old group has a significantly lower VSC than every other age group.

(6) If a head of household is married or, especially, separated, then the VSC is significantly higher than if single. It can also be shown that the separated group's VSC is significantly higher than that of the married group.

(7) Central Office groups 2 and 3 are significantly different from group 1 (and from each other). Note that the ranking of their coefficients is the same as for the computed monthly bill.

(8) The age/years-in-Chicago interaction is significant. Its implications are explained above.

9.5 Conclusions

Several conclusions can be drawn from the analyses which have been completed.

(1) The customer-to-customer variability is much larger than the month-to-month variability.

(2) Demographic characteristics of households, although having a significant influence on the telephone bills, explain only 24 percent of the total customer-to-customer variability of bills and 19 percent of that of vertical service charges. Although this is reasonable for a cross-section study, it is not high enough to allow for accurate prediction of individual customer bills. We may, however, still be able to predict aggregates with good accuracy.

(3) We find computed monthly bill is a monotonic function that (a) increases with increasing income, (b) decreases with increasing age of the head of household, and (c) increases with an increasing number of household members who are at least 10 years of age for both white and other customers. Customers living in Chicago over 15 years have a significantly lower bill than customers living in Chicago for a shorter period of time. There is also a strong race effect, with black customers having a significantly higher coefficient than white and other customers in households with four or fewer members at least 10 years of age.

(4) For vertical service charge, several demographic subgroups are significant. The vertical service charge is higher for households that (a) have a high income, (b) are black, (c) have heads who have lived in Chicago over 15 years, if they are under 45 years of age, (d) have two members rather than one member at least 10 years of age, (e) have heads of age 25 to 34 rather than over 44, or of age 35 to 44 rather than over 64, if they have lived in Chicago over 15 years, (f) have heads of age under 25 rather than over, if they have lived in Chicago 1 to 15 years, (g) have heads who are separated rather than either single or married, or who are married rather than single.

(5) Factors other than the available demographics affect both the computed monthly bill and vertical service charge. This is seen by the amount of unexplained variation in the models and the statistical significance of the Central Office groupings.

Appendix 9.A

Plots for Computed Monthly Bill and for Vertical Service Charge

Figure Number	Title	Page
9.A.1	Normal Probability Plot of Residuals for -1/Square Root of Bill	293
9.A.2	Residual Plot for -1/Square Root of Bill	294
9.A.3	Residual Plot for Vertical Service Charge in Dollars	295
9.A.4	Fitted Values vs. Expected Values for Vertical Service Charge in Dollars	296

Figure 9.A.1

Normal Probability Plot of Residuals
for −1/Square Root of Bill

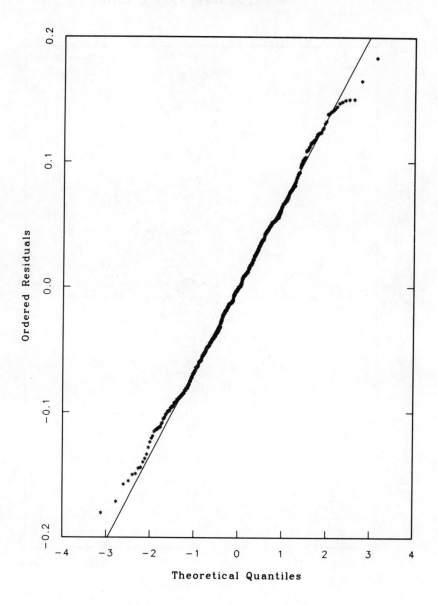

294 *The Effect of Demographics on Telephone Usage*

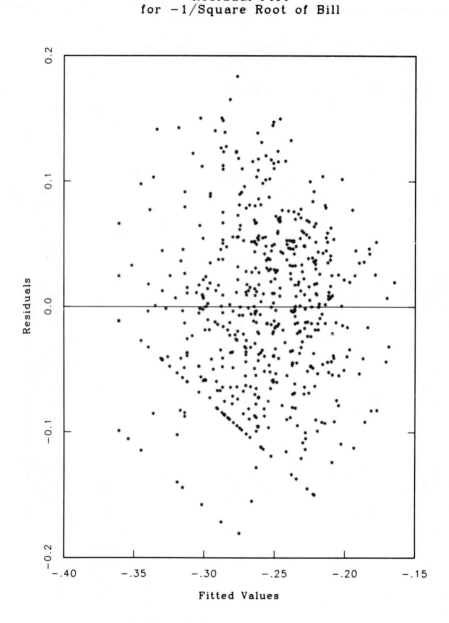

Figure 9.A.2

Residual Plot
for −1/Square Root of Bill

Figure 9.A.3

Residual Plot
for Vertical Service Charge in Dollars

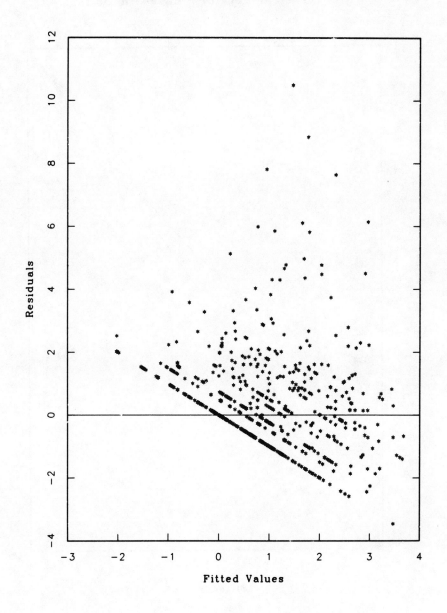

296 *The Effect of Demographics on Telephone Usage*

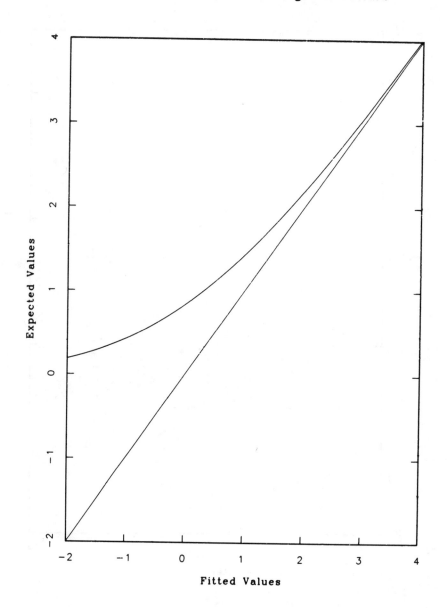

Figure 9.A.4

Fitted Values vs. Expected Values
for Vertical Service Charge in Dollars

Appendix 9.B

Regression Results for Computed Monthly Bill and for Vertical Service Charge

Figure Number	Title	Page
9.B.1	Regression Results for Computed Monthly Bill	298
9.B.2	Extended Tobit Regression Results for Vertical Service Charge	300

Figure 9.B.1

Regression Results for Computed Monthly Bill

R^2	0.240
Residual standard error	0.067
Regression degrees of freedom	25
Residual degrees of freedom	519
F-statistic	6.55

Customer Characteristic	Subgroup	Subgroup Size	Coefficient and Standard Error	t-Statistic
	Intercept		**-0.2794**	-17.47
Household income	5-9 thousand	108	0.0090 (0.0092)	0.98
	9-15 thousand	150	0.0105 (0.0095)	1.11
	15-20 thousand	61	**0.0211** (0.0117)	1.80
	Over 20 thousand	40	**0.0401** (0.0132)	3.03
	No answer	79	**0.0206** (0.0102)	2.01
Racial group	Black	219	**0.0582** (0.0119)	4.89
	Other	47	0.0050 (0.0104)	0.48
Years lived in in Chicago	Over 15	424	**-0.0253** (0.0073)	-3.45
Number of household members at least 10 years of age	Two	223	0.0156 (0.0096)	1.62
	Three	75	**0.0439** (0.0121)	3.63
	Four	45	**0.0512** (0.0150)	3.42
	Five	39	**0.0530** (0.0168)	3.16
	More than five	35	**0.0698** (0.0180)	3.88

Figure 9.B.1 (cont.)

Customer Characteristic	Subgroup	Subgroup Size	Coefficient and Standard Error	t-Statistic
Age the head of household	25-34	163	-0.0070 (0.0119)	-0.59
	35-44	102	-0.0189 (0.0130)	-1.45
	45-54	101	-0.0207 (0.0129)	-1.60
	55-64	81	**-0.0338** (0.0132)	-2.56
	65 or older	61	**-0.0492** (0.0142)	-3.46
Marital status	Married	314	-0.0012 (0.0100)	-0.12
	Separated	41	0.0184 (0.0130)	1.42
	Divorced	45	0.0109 (0.0124)	0.88
	Widowed	58	-0.0065 (0.0123)	-0.53
Central Office group	Group 2	96	**0.0370** (0.0086)	4.28
	Group 3	111	0.0123 (0.0079)	1.56
Interaction of black households and household size		219	**-0.0097** (0.0040)	-2.45

Figure 9.B.2

Extended Tobit Regression Results for Vertical Service Charge

R^2 0.186
Residual standard error 1.56
Standard deviation of latent error 2.019

Customer Characteristic	Subgroup	Subgroup Size	Coefficient and Standard Error	z-Statistic
	Intercept		-2.146	
Household income	5-9 thousand	108	**0.530** (0.227)	2.34
	9-15 thousand	150	**0.694** (0.230)	3.01
	15-20 thousand	61	**1.018** (0.423)	2.41
	Over 20 thousand	40	**1.007** (0.601)	1.68
	No answer	79	**0.786** (0.317)	2.48
Racial group	Black	219	**1.210** (0.058)	20.85
	Other	47	0.598 (0.494)	1.21
Years lived in Chicago	Over 15	424	**1.428** (0.0739)	19.32
Number of household members at least 10 years of age	Two	223	**0.488** (0.138)	3.53
	Three	75	0.827 (0.526)	1.57
	Four	45	0.751 (0.696)	1.08
	Five	39	0.679 (0.694)	0.98
	More than five	35	-0.072 (0.618)	-0.12

Figure 9.B.2 (cont.)

Customer Characteristic	Subgroup	Subgroup Size	Coefficient and Standard Error	z-Statistic
Age of the head of household	25-34	163	0.216 (0.698)	0.31
	35-44	102	0.332 (0.702)	0.47
	45-54	101	-0.873 (0.706)	-1.24
	55-64	81	-0.714 (0.711)	-1.00
	65 or older	61	-1.210 (0.715)	-1.76
Marital status	Married	314	**0.300** (0.137)	2.19
	Separated	41	**1.459** (0.449)	3.25
	Divorced	45	0.332 (0.504)	0.66
	Widowed	58	-0.040 (0.597)	-0.07
Central Office group	Group 2	96	**1.508** (0.268)	5.63
	Group 3	111	**0.676** (0.211)	3.20
Interaction of 1-15 years in Chicago and age		424	**0.452** (0.035)	12.77

CHAPTER 10

Toll Usage

Susan J. Devlin and I. Lester Patterson

10.1 Introduction

This chapter considers toll calls, a service for which usage sensitive pricing already exists. The customer usage measures examined are the number of calls, total conversation time, average duration, and total toll charges. These usage measures are related to four demographic characteristics — race of the head of household, family income, age of the head of household and years of residence in Chicago — considering one of these characteristics at a time.

Two graphical techniques are utilized — bar graphs, to provide simple summaries of the number of customers in various sub-samples, and the notched box plot, a technique described in Chapter 5 for comparing empirical distributions of several sets of data.

In Section 10.2 the data are described. In Section 10.3 the toll usage data are analyzed. In Section 10.4 comparisons are made between the toll analyses and comparable analyses for local and suburban usage. Finally Section 10.5 summarizes the findings and makes recommendations for possible directions in which the toll study might continue.

10.2 The Data

Let us define a "toll call" as a completed, individually billed call between Chicago and any point that is outside the local and suburban calling area. We exclude credit card and third-party calls, so that the telephone number billed belongs to one of the telephones involved in the call. Toll calls here include both intrastate and interstate calls, but exclude calling attempts that are incomplete. "Total toll charges" are analogously defined.

Individual calling and billing information was collected for most of the sample customers for several points in time. The April, 1974, data are used here since toll, suburban and local individual calling information is available for most sampled customers.

Several problems with the April, 1974, data already have been discussed in Chapter 3. An additional problem which becomes relevant in this chapter is that customers who resided in Chicago less than one year as of April, 1974, are not represented in the sample calling data. No correction was made for these problems.

About 15 percent of the customers' bills were missing or incomplete. The customers with missing bills were deleted from the sample where necessary. Though basically the data available seemed to be in reasonably good condition, three customers whose toll usage was among the greatest had portions of their toll detail missing. In checking this more carefully, it was also found that some of their questionnaire responses appear to be incorrect. In one case the customer appears to be a business in the home. For these reasons the three were deleted from the sample. However, since they represent 3 of the 5 customers with toll usage over $100 in the month, approximation of their usage was made, and all results were checked to see if including these approximations would significantly affect the analysis or conclusions.

After correcting the data and eliminating customers with errors affecting toll analyses, 465 questionnaire respondents were left in the Chicago sample.[1] From this a subpopulation was defined as those respondents who were billed for at least one toll call as indicated in the April, 1974, data. About 50 percent of the respondents made at least one toll call, leaving 233 customers in the toll sub-sample and 232 in the no-toll sub-sample.

---

[1] A few sample customers are telephone company employees who receive partial "concessions" on certain segments of their telephone bills, although not on toll. Such customers are included in this study.

For Section 10.4 below, where comparisons are made with local and suburban usage patterns, additional customers were deleted for whom local and suburban calling detail was unavailable. Here the full sample was reduced to 400 customers, of which 338 (85 percent) made at least one suburban call. All of the customers made at least one local call.

10.3 Description of the Toll Findings

Four types of toll usage are considered in this study. They are the number of toll calls, total toll conversation time, average duration of toll calls, and total toll charges. How these usage measures vary, as related (in a univariate sense) to the customer's race, income, years lived in Chicago and age of the head of household, will be described in this section.

10.3.1 Number of Customers — Bar Graphs

First look at who the toll customers are. The bar graphs displayed in Figures 10.A.1 and 10.A.2 explore how representative the toll sample is of the full sample.

Figure 10.A.1a relates to race. The plain bars indicate the percentage representation of each racial group within the toll sub-sample. The corresponding percentages of customers who made no toll calls are represented by the hashed bars. For example about 42 percent of the respondents to the questionnaire who make at least one toll call are black; about 30 percent of the no-toll customers are black. Those customers whose race was not indicated are not counted. Using a χ^2 test of homogeneity of parallel samples, the differences between the toll and no-toll percentages are significant at the 5 percent level, though not significant at the 1 percent level. (The χ^2 statistic with 3 degrees of freedom is 10.7.) Ignoring the other racial categories, the difference between the white and black sub-samples is also significant. This difference suggests that black customers are more likely to make toll calls than white customers in Chicago.

Figure 10.A.1b shows the percentage of the toll and no-toll sub-samples in several family income categories. The χ^2 test indicates that the differences in the income category percentages are not significant. (The χ^2 statistic with 3 degrees of freedom is 2.8.)

Figure 10.A.2a compares the two sub-samples relative to the number of years which heads of households have lived in Chicago. There is an 18 percentage point drop in the proportion of toll customers residing in Chicago over 15 years (72 percent) from the no-toll customer sample (90 percent). The reverse is seen for categories corresponding to under 15 years of residence, in which case there is an over representation of customers in the toll sub-sample. These differences between toll and no-toll are significant at the 0.5 percent level. (The χ^2 statistic with 2 degrees of freedom is 24.1.) Clearly, proportionally fewer customers who have resided in Chicago a long time make toll calls, and thus short-term customers are more likely to make toll calls.

Figure 10.A.2b relates to the age of the head of the household. The patterns for toll and no-toll are not significantly different. (The χ^2 statistic with 5 degrees of freedom is 5.7.) There is a pattern to the differences displayed; proportionally more customers over 45 in the sample do not make any toll calls. Since the χ^2 statistic does not consider the sequential nature of the age categories, it may not be sensitive enough to test for the significance of this slight though consistent pattern.

From Figures 10.A.1 and 10.A.2 it appears that the subset of toll customers differs substantially from Chicago customers as a whole in terms of demographic makeup. Next, how toll usage within this subset varies by demographic characteristics will be studied.

10.3.2 Number of Toll Calls

Figure 10.A.3 is a set of box plots which displays the distribution of the number of toll calls per month for various racial groups. The race categories on the plot are white, black, Spanish, other and NA. The last category, NA (no answer), is comprised of the customers who answered some of the questionnaire but did not indicate their race.

First the definition of a box plot is reviewed. Recall that the horizontal dashed line in the middle of the box for each group indicates the median. The value of the median is written above this dashed line. A box extends from the lower quartile to the upper quartile, with lines drawn from the box to the extremes of the data. When two sets of data are compared, if the notches of the two boxes do not overlap, this suggests that the medians are significantly different at approximately the 5 percent level. When the hypothesis is considered that more than two groups have the same median, wider notches would be required to maintain an overall 5 percent significant level. The last three boxes on Figure 10.A.3 are dashed, indicating that the notches around the median extend beyond one of the two quartiles. Very often this will occur when the group size is small or when the distribution is skewed.

Two things should be kept in mind regarding the box plots presented in the remainder of this chapter. First, when several notches overlap, the wider notches for multiple comparison would necessarily overlap, thus the median differences would be statistically non-significant in a multiple comparison sense. Second, not only because the notches provide an approximate test, but also because the significance level is arbitrarily chosen to be 5 percent, the notches are viewed and interpreted by the authors as rough guides, and the notches are not used to perform formal hypothesis tests. No additional formal tests are referred to.

The width of each box is proportional to the square root of group sample size. The number in each race group is indicated below each box. For example the width of the box for white customers in Figure 10.A.3 is much larger than the width of the box for Spanish customers, as there are 111 customers in the white group versus 15 customers in the Spanish group.

Though the median number of calls for black customers (3.0 calls) is greater than that for white customers (2.0 calls), the notches about the median of each box in Figure 10.A.3 overlap greatly. Thus, there is no significant difference between the medians of the racial groups, when one considers the number of toll calls per customer.

The no-answer group deserves consideration due to the bias introduced if those customers are not random representatives of the population. An extreme case might be if all of the no-answer group really belonged in one racial group. As an experiment, the box plot for the black customers was modified to include the NA group. This experiment was repeated to consider the effect on each of the other racial groups. Though the boxes changed slightly in some cases, the changes were not sufficient to reverse the above observations. Even if only the extreme NA observations were included in the black group, the median does not increase, and the notches widen only slightly. The three customers with large toll bills who were eliminated due to missing toll detail and dubious questionnaire data were identified as two white customers and one customer in the "other" category. Including them would not substantially change Figure 10.A.3. The most noticeable effect would be to increase the white median to 3.0. In no plot presented in this chapter would including these three customers alter conclusions, thus they will not be referred to in any discussions of subsequent plots.

The medians of some of the box plots are much closer to the bottom quartile than to the top quartile, and the range of the top 25 percent is long as compared to the bottom 25 percent. This suggests that the data are skewed and thus non-Normal. As the notches and significance level are defined assuming Normality in the middle of the distribution, this suggests that perhaps the data should be transformed, so that the distribution is more nearly Normal before the box plots are made as is done in Chapters 5 and 7. This was done in several of the cases that are presented in this section using the square root and log transformations. In no case did this reverse the conclusion drawn from looking at the original data. This empirical evidence confirms what McGill *et al.* (1976) said; the notches are approximately correct for many alternatives to Normality. Thus no results using transformations of the data are presented here.

The arrows at the top of the plot indicate that the upper extremes of the corresponding box plots are outside of the plotting range. The value of the extreme is printed beside the arrow. Notice that the largest number of calls per customer is 40 and that this customer is in the black group. If the three deleted customers were included, then the extreme would be a white customer with about 75 toll calls.

In conclusion, Figure 10.A.3 suggests that for the toll sub-sample the distribution of the number of toll calls does not significantly differ between black and white customers. The sample sizes for the other groups are too small to generate confidence about the results.

Now the toll customers are grouped by 1972 family income to see if it differentiates among toll calling patterns. The numbers of toll calls versus family income are displayed in Figure 10.A.4. The income categories are defined differently than in Chapter 5, as preliminary analysis of the toll data indicated that grouping of incomes into the categories under $9,000, $9,000 to $19,999, and $20,000 or more, is more informative for the analysis presented here. Clearly the first two boxes in Figure 10.A.4 reflect similar distributions, with both medians at two toll calls. For the highest-income category the median is five toll calls. The notches indicate that the difference in medians between the highest-income group and either of the lower-income groups is significant at approximately the 5 percent level. Even if all those who did not give their income were classified as having incomes of $20,000 or more, the median would only be pulled down to four toll calls, and the modified notches would still not overlap with either of the under-$20,000 income categories.

As previously indicated in Figure 10.A.3, the highest number of toll calls was 40, and it occurred in the black group. Now in Figure

10.A.4 the same black customer with 40 toll calls appears to have income between $9,000 and $19,999.

In conclusion, Figures 10.A.4 and 10.A.1b demonstrate a clear association between family income and how many toll calls a customer is likely to make.

Next the customers are grouped by the number of years lived in Chicago to see if the calling patterns will vary based on this factor. See Figure 10.A.5. As only one customer was a non-respondent for years lived in Chicago, no NA category is included. The medians are insignificantly different, as the notches for each pair of groups overlap. The slight, though insignificant, downward pattern in the medians is supported by a downward trend in the quartiles and by Figure 10.A.2a, which suggests that long term residents are less likely to make any toll calls. Unfortunately the 166 customers in the greater-than-15-years group cannot be sub-divided to see if the pattern continues. Also any conjectures made from the data excludes very short term residents (under 1 year), due to the one-year delay from the collection of questionnaire data to the collection of toll calling detail.

A final customer characteristic considered as a calling pattern discriminator is the age of the head of household. Box plots of the number of toll calls for six age groups are displayed in Figure 10.A.6. No patterns in the boxes suggest a way of differentiating calling patterns by age of the head of household. Recall that there was also no strong evidence of a dependence of toll usage with age from comparing the toll and no-toll sub-samples.

10.3.3 Total Toll Conversation Time

The number of toll calls indicates how often the telephone line is used for toll calls, whereas total conversation time (TC) measures the time the line was recorded as making toll calls.[2] Thus a customer billed for 10 one-minute calls (TC=10) has the same TC-value as a customer who makes 1 ten-minute call (TC=10) but a different value from a third customer who makes 10 calls of 10 minutes each (TC=100). Total toll conversation time per toll customer is compared to the same

---

[2] Total recorded conversation time is a biased estimate of actual conversation time. This will be discussed more fully in the next section. Here total conversation time refers to the recorded time. Note that the recorded time is equivalent to the billed time except that the former excludes any unutilized portion of any initial period.

310   *The Effect of Demographics on Telephone Usage*

four demographic measures as in the previous sub-section: race, income, years lived in Chicago, and age of the head of household.

First consider total conversation time versus race, which is displayed in Figure 10.A.7. Since the black and white races are the largest samples, they are of primary concern. Though the notches suggest that there is no statistically significant difference between these medians, the 11-minute difference between the white and black group medians is interesting. Notice that the upper extremes for both races, black and white, are beyond the maximum plotting range of 200 minutes. One black customer had a total conversation time as large as 393 minutes, and a white customer had as many as 434 minutes of total conversation time.

Total conversation time versus family income is displayed in Figure 10.A.8. For customers whose family income is $20,000 or more, the median length of time spent in toll conversations (80 minutes) is about three times that of either of the lower income categories. This difference appears to be statistically significant. Though the median total conversation time for customers whose family income is less than $9,000 is larger than that for customers in income category $9,000 to $19,999, this difference is small and insignificant. The dip in the upper quartile for income group $9,000 to $19,999 suggests that total conversation time for customers in this income category is more tightly distributed about the median.

Figure 10.A.9 gives some indication that total conversation time for a short-term Chicago resident (10 years or less) is likely to be greater than for a long-term resident (greater than 15 years); however, the notches about the medians overlap.

The distributions of the total conversation time per customer versus age are displayed in Figure 10.A.10. Though there may be a slight upward trend in the medians and upper quartiles for the first four age groups, the notches are wide. No pattern in the boxes strongly suggests a way of differentiating calling patterns by age of the head of household.

10.3.4 Average Duration of Toll Calls

In this section the average duration of a toll conversation (*i.e.*, TC/number of toll calls) per customer will be discussed. Number of calls and total conversation time have already been discussed. Now average duration will be looked at to see if it is more strongly related to household characteristics.

Recorded conversation durations for toll data, as well as recorded conversation durations for local and suburban, are biased estimates of actual conversation durations due to a rounding rule defined in Section 5.4.2. In that section a procedure was described which attempts to estimate this bias and adjust conversation lengths accordingly. For the analyses described in this chapter no such adjustment was used. The sparsity of toll data does not allow assessing the fit of a distribution. Several studies have shown that the log Normal is a good approximation to the distribution of length of conversation of toll calls aggregated over customers, but that information is not sufficient to indicate the appropriate probability distribution for individual customers. The present authors prefer to use recorded conversation lengths, avoiding additional assumptions, for direct comparisons between types of calls, as is done in Section 10.4 below. Also, they feel that these are the values upon which pricing is based.

First let us consider average duration versus race, which is displayed in Figure 10.A.11. Though the median for black customers is one minute greater than that for white customers, the notches around the medians overlap. The interquartile range (a measure of variability) of the average duration for black customers is nearly twice that of white customers. Again neither the Spanish nor "other" group appears to be different from the others.

Average duration versus family income is displayed in Figure 10.A.12. The overlapping notches indicate that there is no significant association between average duration and family income. The median average duration for the over-$20,000 income category is larger than the other medians. This may explain why the corresponding difference for total conversation time (Figure 10.A.8) was proportionally greater than that for the number of toll calls (Figure 10.A.4).

Average duration versus years lived in Chicago is displayed in Figure 10.A.13. The average duration of a toll call for a customer appears not to be related to the number of years a customer lived in Chicago.

In Figure 10.A.14 the age of the head of household is not seen as a factor affecting the distribution of average duration.

10.3.5 Toll Charges

Total toll charge is different from the other toll usage measures in that it uses more information about calls, as the charge for a call is based on information such as terminating location, time of day, and type of call. It may be that these factors which affect the charge differentiate between customer groups. For example some group could be calling longer distances, making greater use of time-of-day discounts, *etc.* The toll charge is probably the most visible effect to customers of their toll usage and is of great concern in planning.

First consider toll charges billed in April, 1974, by race, which are displayed in Figure 10.A.15. Though the notches overlap, the median for black customers is 40 percent greater than the median for white customers. This difference presumably is affected by the greater number and average duration of toll calls for black customers than for white customers. Due to the relative size of the differences and due to the highly visible nature of the total charges to the public, further analysis, either correcting for several demographic variables simultaneously or considering the factors contributing to price differentials, needs to be done to explain this potential difference between the racial groups.

Another customer characteristic to be considered as a calling pattern discriminator is income. The distribution of the toll charges per customer versus income, displayed in Figure 10.A.16, reflect results already seen. The medians for the first two income categories are both about $5.00. Customers in the second ($9,000 to $19,999) group are more tightly distributed about the median than in the first. The median toll charges ($15.19) for customers with family incomes of $20,000 or more is about three times that for the lower income categories. This large difference is statistically significant.

Now let us consider years lived in Chicago to see what association it may have with toll charges (see Figure 10.A.17). The medians decrease as years of residence increase. The notched interval for the last group just barely overlaps with that of the first group. Notches corresponding to a 10 percent significance level would not have overlapped. The tops of the boxes also show a decreasing trend in the toll charges as the number of years lived in Chicago increases. In conclusion, the longer term Chicago residents are likely to spend less money on toll calls. This is consistent with the implications of other usage measures, though no significant results were seen for them.

Next Figure 10.A.18 considers age to see what association it may have with toll charges. Again age seems not be be a discriminator among calling patterns.

In the next section a comparison will be made to see if toll calling patterns are similar to local and suburban calling patterns.

10.4 Toll vs. Local and Suburban Calling — Demographic Comparisons

Chapter 5 compared the distribution of local and suburban calling patterns (number of calls, total conversation time and average duration) among different race, income and age groups. The local and suburban usage data were collected at the same time as the toll data were collected, April, 1974. Comparisons among customers grouped by the number of years lived in Chicago were not reported. In this section the local and suburban findings are reviewed, using variable-width notched box plots. The corresponding local, suburban and toll analyses are compared. Toll charges cannot be considered as there is no direct local usage charge to compare it against.

The local and suburban samples are defined with slightly different data restrictions, thus sample sizes differ slightly from previous analyses presented. Demographic groupings are slightly different from those in Chapter 5 for direct comparison to toll analysis. Furthermore, in reviewing suburban usage, only customers who made at least one suburban call are considered here. Finally, when considering total conversation time and average duration, recorded conversation durations are used for comparison to toll results. For the above reasons some local and suburban box plots, discussed in Chapter 5, are reviewed and re-presented in a modified form. The inferences made for these local and suburban box plots are consistent with what was seen in Chapter 5.

For those customers for whom local and suburban calling detail were available, every customer made at least one local call in the April, 1974, billing period. Overall far more customers make suburban calls (85 percent) than make toll calls (50 percent). Figures 10.A.19 and 10.A.20 show the proportional breakdown by demographic groups of customers who made at least one suburban call and of customers who made no suburban calls. The suburban and no-suburban strata can be compared to the toll and no-toll strata by comparing Figures 10.A.19 and 10.A.20 to Figures 10.A.1 and 10.A.2.

The difference in the racial breakdown between the suburban and no-suburban sub-samples is greater than the difference seen for the toll and no-toll sub-samples. Furthermore the pattern is reversed. Whereas Figure 10.A.1a implies that black customers are significantly more likely to make toll calls than white customers, Figure 10.A.19a suggests that white customers are more likely to make suburban calls than black customers are. (The suburban/no-suburban χ^2 statistic with 3 degrees of freedom is 33.4.)

314 *The Effect of Demographics on Telephone Usage*

The proportional breakdown of customers by family income is similar for suburban and toll (Figures 10.A.1b and 10.A.19b). Proportionally more customers from the larger-income categories make toll and suburban calls, although this difference is not significant at the 10 percent level for either suburban or toll calls. (The suburban/no-suburban χ^2 statistic with 2 degrees of freedom is 2.7.)

Whereas Figure 10.A.2a implies that customers who have resided in Chicago over 15 years are less likely to make any toll calls, Figure 10.A.20a shows no difference between suburban callers and no-suburban callers with respect to the years of residence in Chicago. (The suburban/no-suburban χ^2 statistic with 2 degrees of freedom is 0.9.)

The proportional breakdown of customers by age is not significantly different between suburban and no-suburban customers, just as was seen for toll and no-toll. (The suburban/no-suburban χ^2 statistic with 5 degrees of freedom is 5.1.) However the pattern of the differences displayed for toll/no-toll in Figure 10.A.2b does not occur in Figure 10.A.20b for suburban/no-suburban.

Next toll, local and suburban calling are compared using three measures of usage — the number of calls, total conversation time and average duration. As additional information was not gained from total conversation time, no plots involving this measure are included. The discussion is organized by demographic characteristic.

10.4.1 Race

The box plots in Figures 10.A.21 and 10.A.22 compare, for different racial groups, the distribution of the number of local calls per customer and average duration per customer for local calls, respectively. These figures show that the median number of local calls made by black customers is significantly higher than that made by white customers. The median for black customers of average duration of local calls (6.5 minutes) is over one minute longer than that for white customers (5.1 minutes). In Figure 10.A.22 the variability for the Spanish group does not result in very large notches. The median average duration of local calls for Spanish customers is significantly shorter than that for black customers.

The box plots in Figures 10.A.23 and 10.A.24 are the corresponding plots describing suburban usage for those customers who make at least one suburban call. The patterns are reversed from that seen for local calling. The median number of calls for white customers (10.5 calls) is over 3 times that for black customers (3.0 calls). The median for Spanish customers (9.0 calls) is significantly greater than that for black customers. The average duration for white customers (6.0 minutes) is twice that for black customers (3.0 minutes).

These noticeably different local and suburban calling patterns between black and white customers are further contrasted with the toll calling patterns seen in Figures 10.A.3 and 10.A.11: Calling patterns for black toll customers are not significantly different from that of white toll customers.

One might have hypothesized that suburban calling patterns were like short-haul toll; however, the box plots suggest that suburban and toll calling are quite different. A more likely hypothesis is that the critical factors are the demographic characteristics of the area to where calls are made and that there is a great difference between racial groups as to the inclination to make calls to areas with different demographic compositions. One proxy for such a "community of interest" measure might be whether the called area is a city, suburb, or rural area. To consider this more carefully, usage would need to be separated by mileage groups and/or by area codes of the terminating numbers.

The charges for toll calls are typically greater than for local or suburban calls, and the charge is more visible with the individual billing of toll calls. Furthermore, it is likely that any community of interest decreases with distance. For these reasons, it is not surprising that far fewer toll calls than local or suburban calls are made. Recall that 50 percent of the customers in the full sample made no toll calls for the period which the data represent. Ignoring suburban calling, the median average duration of a toll call is about three minutes longer than that of a local call for white and black customers.

10.4.2 Income

Figure 10.A.25 shows that as family income increases, there is a slight, though insignificant, increase in the number of local calls per customer. The box plots in Figure 10.A.26 are the corresponding plots describing suburban usage. Figure 10.A.26 shows a significant increase in the median number of suburban calls in the $9,000-to-19,999 income category over that of customers in the under-$9,000 category. The median for the highest income category shows a further increase, though the notches for this small sample overlap the notches for the middle-income category.

The toll usage box plots corresponding to Figures 10.A.25 and 10.A.26 for family income are in Figure 10.A.4. Whereas Figures 10.A.26 and 10.A.4 suggest a significant upward trend in the number of suburban and toll calls with income, for local calls (Figure 10.A.25) there is no significant difference between income levels. Whereas for toll there is no evidence of a difference between the two categories under $20,000, for suburban each group appears to be different from the others.

As the box plots suggest that income does not affect the distribution of average duration for local, suburban or toll calls, the local and suburban plots are not shown.

10.4.3 Years of Residence in Chicago

One might anticipate that the longer one resides in Chicago, the more local calls one might make; whereas the opposite pattern should appear for toll. Figures 10.A.27 and 10.A.28 compare, for groups of different numbers of years of residence in Chicago, the distributions over customers of the numbers of local calls and average durations of local calls, respectively. Figure 10.A.27 shows that the number of local calls is lower for customers living in Chicago ten years or less than for those living in Chicago over 15 years; however, the notches for these two boxes overlap. The distributions of average duration (Figure 10.A.28) do not differ significantly. Of the sample respondents, 80 percent resided in Chicago more than 15 years.

Years of residence in Chicago is a significant factor affecting both measures of suburban usage (Figures 10.A.29 and 10.A.30). Though two of the boxes are dashed in Figure 10.A.29, the notches are defined. The notch for the 10-years-and-under category would not overlap with the notch for the over-15-year category. Clearly long term Chicago residents are likely to make more suburban calls, and the average duration of their suburban calls are longer (Figure 10.A.30). These figures indicate that as a customer's radius of familiarity for an area grows with length of residence, this is reflected in the number of suburban (*i.e.*, more distant) calls made. Recall that this trend was not reflected in Figure 10.A.20a, which compares suburban and no-suburban calling customers.

Compare these local and suburban usage figures with Figures 10.A.5 and 10.A.13 for toll usage versus years of residence in Chicago. Although years of residence do not clearly differentiate between medians for the number of and average duration of toll calls, their combination in the form of total toll conversation time (Figure 10.A.9) does suggest that shorter term residents (less than or equal to 10 years) do have higher median total conversation time than customers having resided in Chicago over 15 years. This is the reverse of the suburban calling pattern in Figure 10.A.29 and in the plots for total suburban conversation time, which are not shown here.

10.4.4 Age of the Head of Household

Figure 10.A.31 considers the number of local calls by age of the head of household. Customers whose household head is over 54 years of age tend to make many fewer local calls than customers in the younger age groups. The notches suggest that this difference may be statistically significant. The medians for the age groups under 55 years are around 90 calls, whereas the medians for the over-54-years groups are more nearly 50 calls. As suggested in Chapter 5, this effect may be at least partly a function of family size, which is reflected in age. The differences among the under-54-years groups do not appear to be significant in Figure 10.A.31.

The distributions of the number of suburban calls (Figure 10.A.32) and of the number of toll calls (Figure 10.A.6) are not affected by the age of the household head. If it is true that family size rather than age itself causes the local calling patterns, then it is likely that minors in a family (reflecting larger family size) are making local calls, which are not individually billed, but are not affecting the number of toll or suburban calls for the family.

The distribution of average duration shows no variation with age of the household head for local, suburban or toll calls, thus local and suburban plots are not shown.

10.5 Conclusions and Recommendations

In studying a sample of Chicago residence customers who were billed for at least one toll call for a specified billing period, customers' toll usage patterns, as measured by the number of calls, total conversation time, average duration or toll charges per customer, were not found to vary significantly among the groupings for race, age of the head of household, or years lived in Chicago. Though the distribution of the number of toll calls did not significantly vary between black and white toll customers, proportionally more black customers make at least one toll call. Also, though racial groupings were not shown to be significant discriminators of toll charges, the large differences in median toll charge among sample groups requires further investigation. Proportionally more short-term residents make at least one toll call, while among toll customers the distribution of the number of toll calls is not affected by years of residence in Chicago. Customers with family incomes over $20,000 make more toll calls. The median toll charge for that group is about 3 times the medians for the lower income categories.

One can clearly conclude from the results shown here that local, suburban and toll are three very different markets; generally, demographic patterns do not carry over from one to another. Where race proved to be a discriminator among local calling patterns, in that black customers tend to make significantly more calls and longer calls than white customers, the distributions of toll usage did not differ significantly between racial groups, though the direction of the difference was the same. But for suburban calling, the pattern was clearly in the opposite direction; black customers tend to make significantly fewer suburban calls than white customers. Perhaps the explanation is in the difference between racial groups in their communities of interest. To pursue this one might consider a breakdown of toll calls by their terminating location. However, substantially more data would probably be needed for this.

Though median toll and suburban usage is greater for the family income category over $20,000, family income has no significant association with local usage. Age of the head of household seems to have no relationship with toll and suburban calling patterns, whereas there was a strong drop in the median number of local calls for customers over 54 years of age. Though one might anticipate that local usage is greater for long-term Chicago residents (greater than 15 years), the tendency is not statistically significant. Long-term Chicago residents make significantly more suburban calls. The opposite pattern is seen for toll, though it is not statistically significant.

In reviewing these results, additional questions and alternative analyses come to mind. Of course other demographic measures could be considered. Since, in a number of cases, differences were large though not significant, perhaps a two-way interaction should be considered, such as race by income. It is likely that sample sizes are not large enough to do this, though further aggregation may help.

It would be interesting to consider the local and suburban usage patterns of just those customers who made toll calls. It might be observed that the median number of local and suburban calls is lower among toll customers than among no-toll customers. If toll calls were a substitute for other usage, then such a pattern would be expected. An alternative is that local and suburban usage is equal or greater among toll customers. Either of these alternatives could be caused by demographic stimuli as yet uncovered.

Toll Usage 319

The large differences in toll charges for income and race groups indicate that further study needs to be made of the individual characteristics of each toll call as they affect the charge, such as where different demographic groups tend to call, what types of service are preferred (person-to-person, credit card, direct dial, *etc.*), and how time-of-day patterns differ among demographic groups. Though toll charges clearly differ among some groups (*e.g.*, Figure 10.A.16), other differences may be masked by the interaction of the above mentioned effects.

Clearly there are additional questions concerning toll usage and its relation to local and suburban calling patterns which could be explored with more data. Also, other components of the bill, such as those considered in Chapter 9, might also display interesting patterns when comparing the toll and no-toll sub-samples.

Appendix 10.A

Toll Calling by Demographic Group

Figure Number	Title	Page
10.A.1	Breakdown of Toll and No-Toll Sub-Samples by Race and by Income	322
10.A.2	Breakdown of Toll and No-Toll Sub-Samples by Years Lived in Chicago and by Age	323
10.A.3	Number of Toll Calls by Race	324
10.A.4	Number of Toll Calls by Income	325
10.A.5	Number of Toll Calls by Years Lived in Chicago	326
10.A.6	Number of Toll Calls by Age of Head of Household	327
10.A.7	Total Toll Conversation Time by Race	328
10.A.8	Total Toll Conversation Time by Income	329
10.A.9	Total Toll Conversation Time by Years Lived in Chicago	330
10.A.10	Total Toll Conversation Time by Age of Head of Household	331
10.A.11	Average Durations of Toll Calls by Race	332
10.A.12	Average Durations of Toll Calls by Income	333
10.A.13	Average Durations of Toll Calls by Years Lived in Chicago	334
10.A.14	Average Durations of Toll Calls by Age of Head of Household	335
10.A.15	Total Toll Charges by Race	336
10.A.16	Total Toll Charges by Income	337
10.A.17	Total Toll Charges by Years Lived in Chicago	338
10.A.18	Total Toll Charges by Age of Head of Household	339
10.A.19	Breakdown of Suburban and No-Suburban Sub-Samples by Race and by Income	340
10.A.20	Breakdown of Suburban and No-Suburban Sub-Samples by Years Lived in Chicago and by Age	341
10.A.21	Number of Local Calls by Race	342
10.A.22	Average Durations of Local Calls by Race	343
10.A.23	Number of Suburban Calls by Race	344
10.A.24	Average Durations of Suburban Calls by Race	345
10.A.25	Number of Local Calls by Income	346
10.A.26	Number of Suburban Calls by Income	347
10.A.27	Number of Local Calls by Years Lived in Chicago	348

10.A.28	Average Durations of Local Calls by Years Lived in Chicago	349
10.A.29	Number of Suburban Calls by Years Lived in Chicago	350
10.A.30	Average Durations of Suburban Calls by Years Lived in Chicago	351
10.A.31	Number of Local Calls by Age of Head of Household	352
10.A.32	Number of Suburban Calls by Age of Head of Household	353

322 *The Effect of Demographics on Telephone Usage*

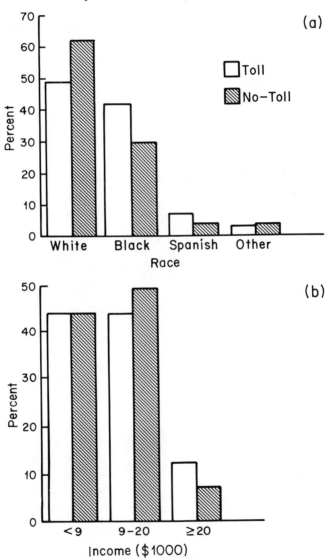

Figure 10.A.1

Breakdown of Toll and No-Toll Sub-Samples by Race and by Income

Figure 10.A.2

Breakdown of Toll and No-Toll Sub-Samples
by Years Lived in Chicago and by Age

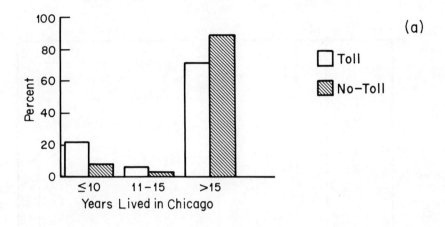

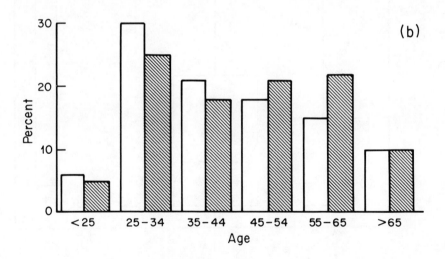

324 *The Effect of Demographics on Telephone Usage*

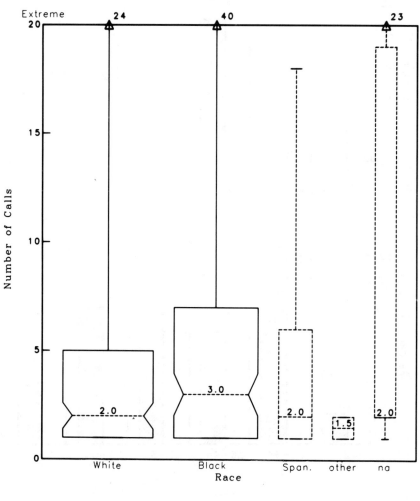

Figure 10.A.3

Number of Toll Calls by Race

Figure 10.A.4

Number of Toll Calls by Income

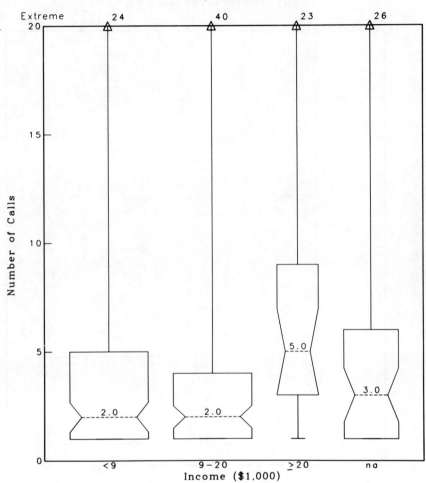

326 *The Effect of Demographics on Telephone Usage*

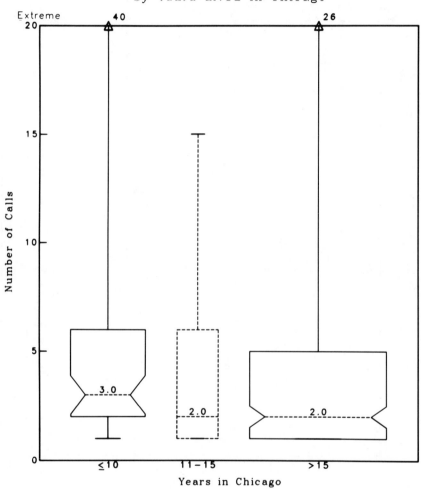

Toll Usage 327

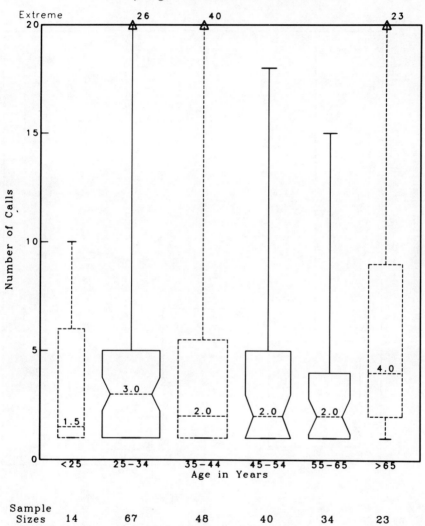

Figure 10.A.6

Number of Toll Calls by Age of Head of Household

328 *The Effect of Demographics on Telephone Usage*

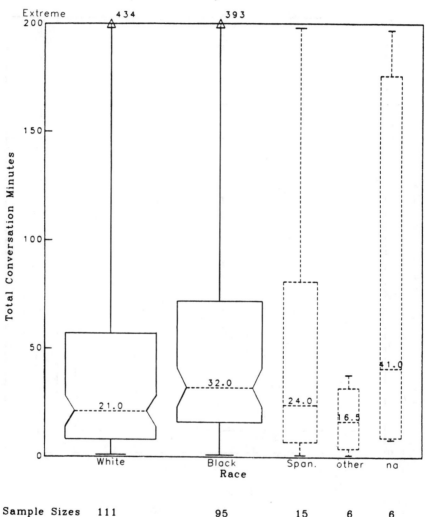

Figure 10.A.7

Total Toll Conversation Time by Race

Toll Usage 329

Figure 10.A.8

Total Toll Conversation Time by Income

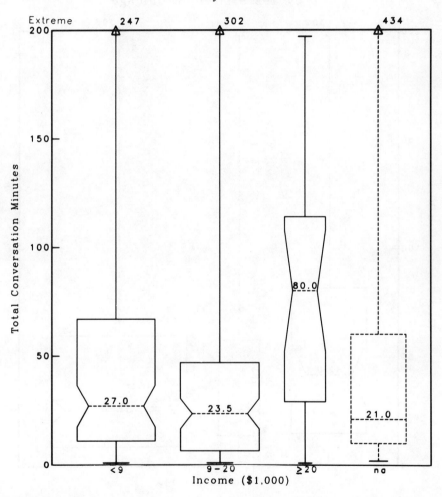

330 *The Effect of Demographics on Telephone Usage*

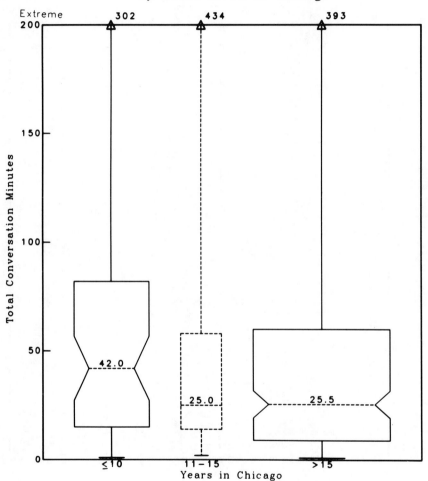

Figure 10.A.9

Total Toll Conversation Time by Years Lived in Chicago

Toll Usage 331

Figure 10.A.10

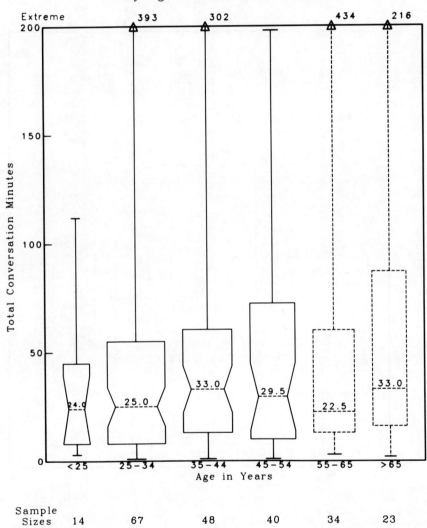

Total Toll Conversation Time by Age of Head of Household

332 *The Effect of Demographics on Telephone Usage*

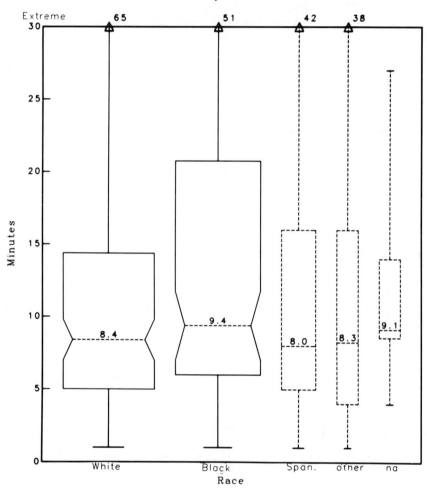

Toll Usage 333

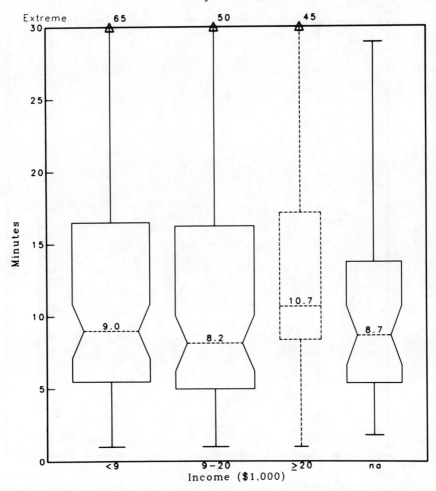

Figure 10.A.12

Average Durations of Toll Calls by Income

The Effect of Demographics on Telephone Usage

Figure 10.A.13

Average Durations of Toll Calls by Years Lived in Chicago

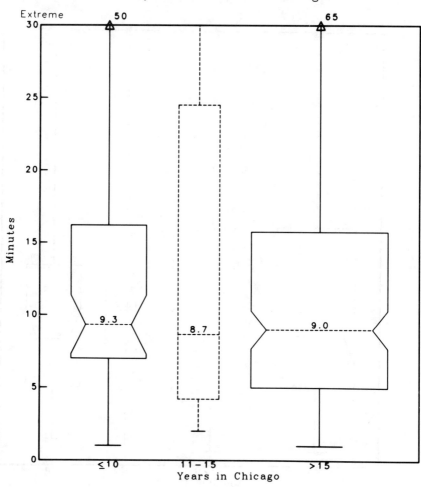

Sample Sizes   51   15   166

Figure 10.A.14

Average Durations of Toll Calls
by Age of Head of Household

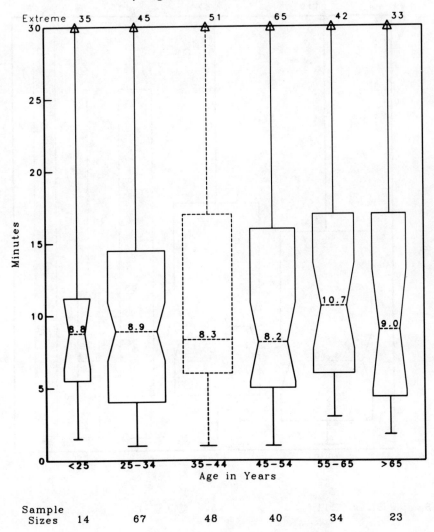

336 *The Effect of Demographics on Telephone Usage*

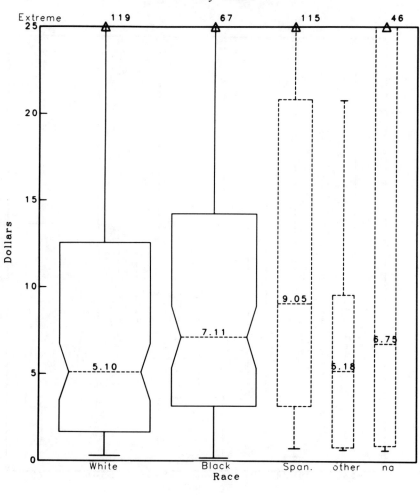

Figure 10.A.15

Total Toll Charges by Race

Toll Usage 337

Figure 10.A.16

Total Toll Charges
by Income

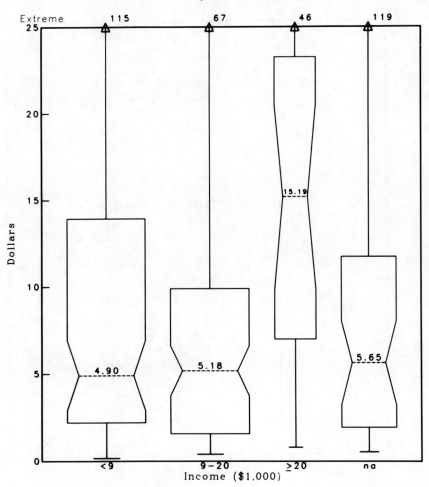

Sample Sizes   85   84   23   41

338 *The Effect of Demographics on Telephone Usage*

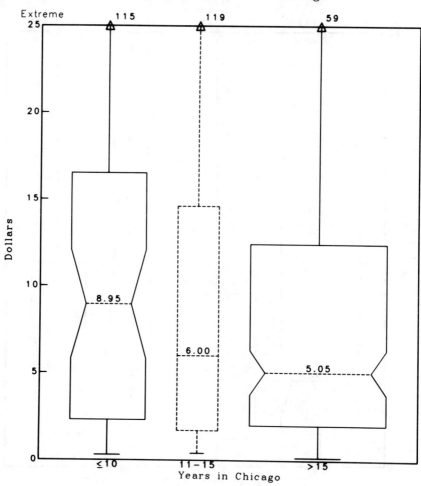

Figure 10.A.17

Total Toll Charges by Years Lived in Chicago

Figure 10.A.18

Total Toll Charges
by Age of Head of Household

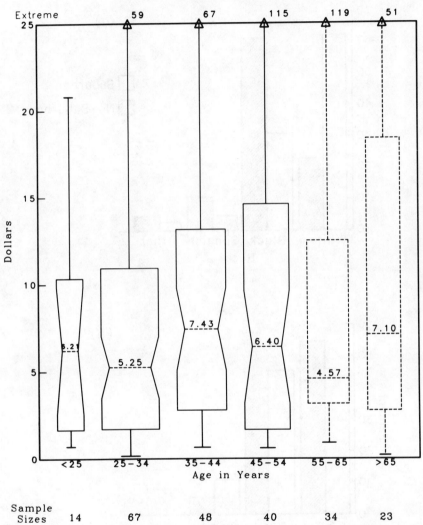

340 *The Effect of Demographics on Telephone Usage*

Figure 10.A.19

Breakdown of Suburban and No-Suburban Sub-Samples by Race and by Income

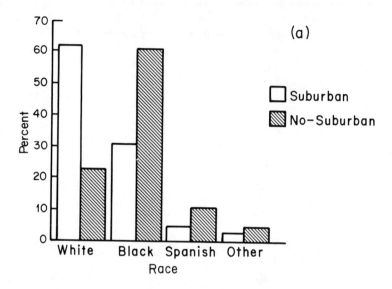

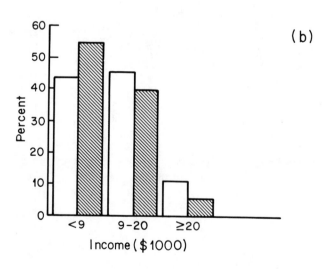

Toll Usage 341

Figure 10.A.20

Breakdown of Suburban and No-Suburban
Sub-Samples by Years Lived in Chicago and by Age

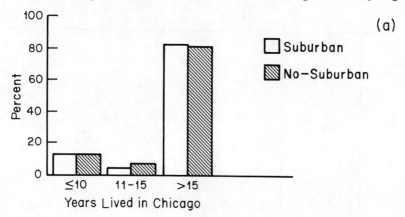

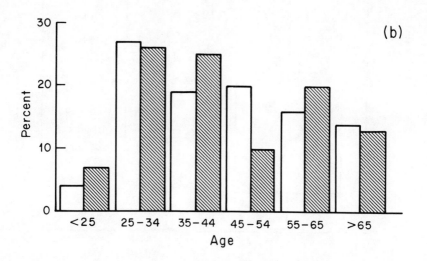

342 *The Effect of Demographics on Telephone Usage*

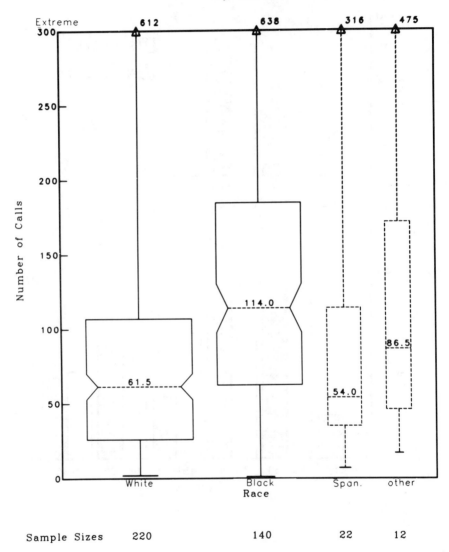

Toll Usage 343

Figure 10.A.22

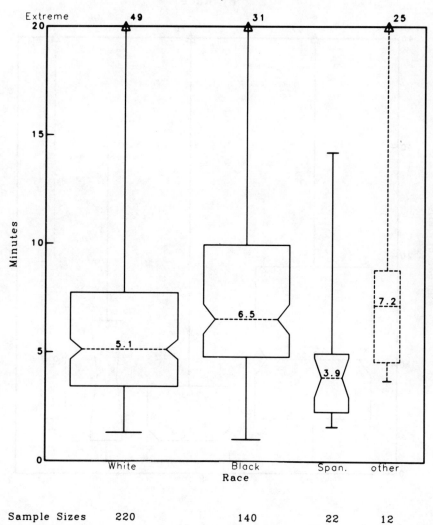

Average Durations of Local Calls by Race

| Sample Sizes | 220 | 140 | 22 | 12 |

344 *The Effect of Demographics on Telephone Usage*

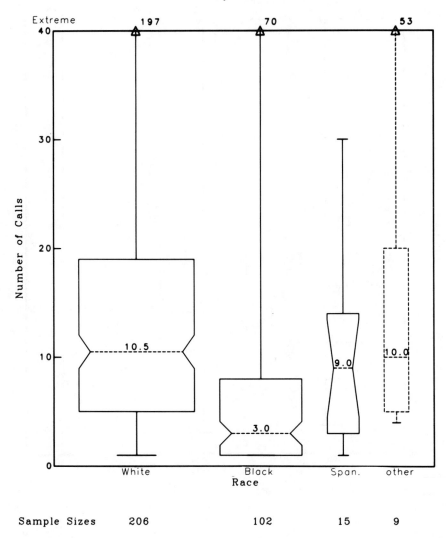

Figure 10.A.23

Number of Suburban Calls by Race

Toll Usage 345

Figure 10.A.24

Average Durations of Suburban Calls by Race

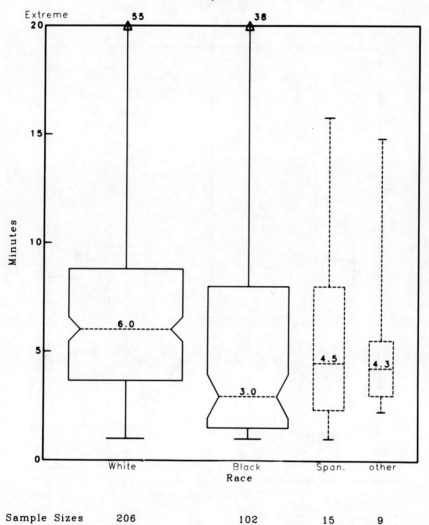

346 *The Effect of Demographics on Telephone Usage*

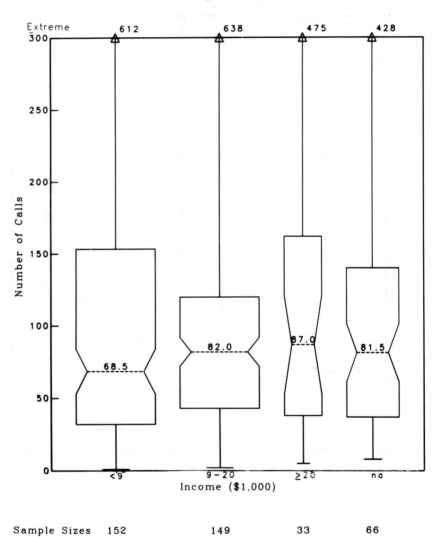

Toll Usage 347

Figure 10.A.26

Number of Suburban Calls by Income

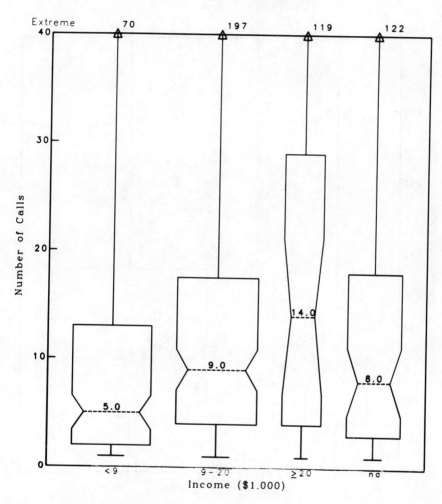

Sample Sizes 123 128 30 57

348 *The Effect of Demographics on Telephone Usage*

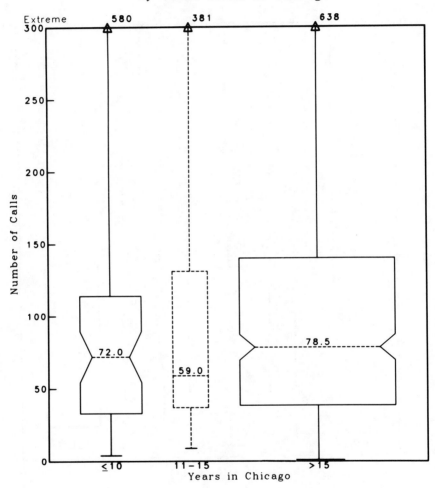

Figure 10.A.27

Number of Local Calls
by Years Lived in Chicago

Toll Usage 349

Figure 10.A.28

Average Durations of Local Calls by Years Lived in Chicago

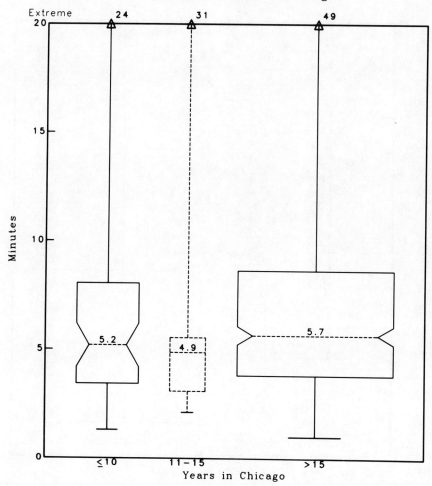

Sample Sizes   53   17   328

350 *The Effect of Demographics on Telephone Usage*

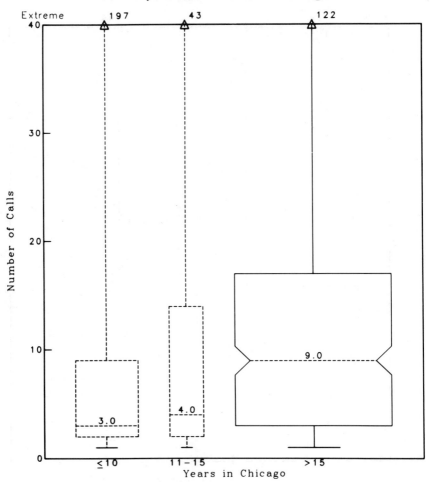

Figure 10.A.29

Number of Suburban Calls by Years Lived in Chicago

Toll Usage 351

Figure 10.A.30

Average Durations of Suburban Calls
by Years Lived in Chicago

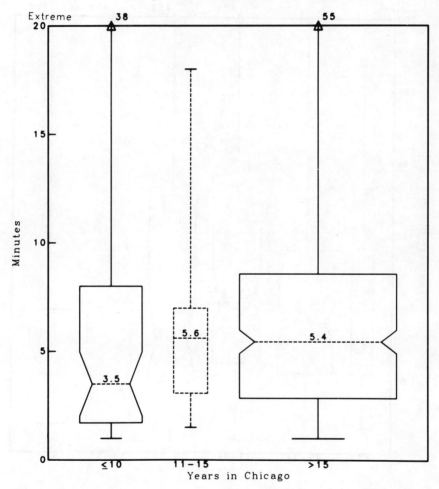

352 *The Effect of Demographics on Telephone Usage*

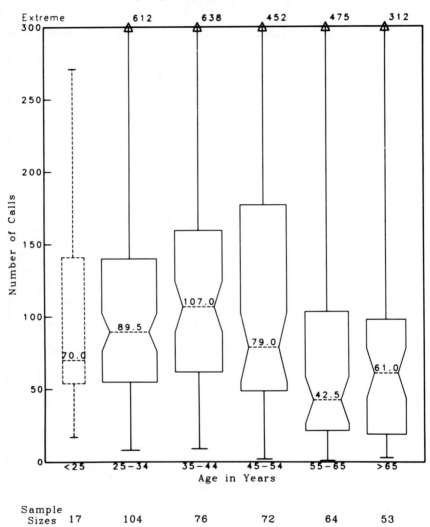

Figure 10.A.32

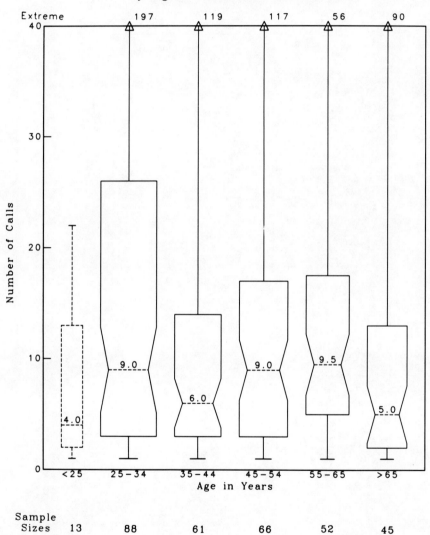

Number of Suburban Calls by Age of Head of Household

CHAPTER 11

Other Studies and Suggestions for Future Research

Belinda B. Brandon, Paul S. Brandon, and Wm. H. Williams

11.1 Introduction

This book has documented a project that has explored the relationship between various measures of the use of telephone services and demographics in Chicago. The emphasis has been on local usage, but results for toll and other services have been reported as well.

The next section of this chapter examines some of the other studies that relate telephone usage to demographic characteristics. Section 11.3 outlines a suggested sample design for a before-and-after study of a pricing change. The following section discusses suggested changes in the demographic questionnaire for any new studies that might be undertaken. The last section of the chapter suggests some new studies that might be carried out to understand better the relationship between telephone usage and demographic characteristics.

11.2 Other Studies

This section of the paper briefly describes five studies that relate demographics to local telephone usage and, when appropriate, compares some of their results to the Chicago study. Two of the studies, using data from Pacific Telephone and Telegraph Co. (PT&T) and from Cincinnati Bell Telephone Co., are essentially completed, although unpublished.[1] The third study, by Southern New England Telephone Co. (SNET), is in progress. PT&T was the first to begin collecting data, but analysis was begun on the Chicago data before any of the others. The Chicago study also has the largest scope. W. J. Infosino at Bell Telephone Laboratories carried out both the PT&T and Cincinnati studies; S. M. Collins of Cincinnati Bell assisted the Cincinnati study. The Management Sciences organization in SNET is carrying out the SNET study. Similar demographic questionnaires were used in the Chicago, Cincinnati, PT&T, and SNET studies.

The Rand Corporation and Technology & Economics, Inc., (T&E) have recently proposed to carry out similar studies. Bridger Mitchell will be the principal investigator in the Rand study, which will be done under a grant from the National Science Foundation. The T&E study will be performed under contract for the Federal Communications Commission.

It may be useful to compare the multiple regression results on the number of local calls from the Chicago, Cincinnati, and PT&T studies (neither the PT&T nor Cincinnati studies deal with measures other than the number of local calls). (1) There is no systematic relationship between income and calling rates for either PT&T or Cincinnati when other demographics are held constant; however, in Chicago the number of calls is higher for the highest-income group than it is for the other groups. This result is not surprising since the studies for the first two areas analyze only customers who, along with the majority of customers in those areas, choose the flat-rate option — *i.e.*, pay a fixed monthly charge for local service regardless of the number of calls — whereas almost all Chicago customers pay a message unit charge for each call. (2) According to all three studies the number of local calls rises as the size of the household increases. (3) In both PT&T and Cincinnati, the number of calls is larger if a household is headed by a female rather than by a male; when the Chicago data are analyzed with a specification identical to that of PT&T and Cincinnati, no such effect is found. (4) For the Chicago and Cincinnati studies, black households have a significantly higher calling rate than white households do; the same effect can be inferred, using Census data, from the PT&T study. (5) For all three studies, households headed by persons 65 years of age or older have a significantly lower calling rate than the others do.

---

[1] A summary of some of the results was presented by Carl Pavarini and William J. Infosino at the Telecommunications Policy Conference, Skytop, Pennsylvania, April 30, 1979.

An intriguing result from the studies of flat-rate customers in PT&T and Cincinnati is that there is an appreciable difference in the calling rates between the studied areas that is not captured by the included demographic effects. A regression estimated from the PT&T data succeeds in predicting the ranking of the calling rates in the Cincinnati Central Office Areas given Cincinnati demographic data, but consistently under-predicts the absolute calling rate. Further, predictions from Cincinnati data over-predict PT&T calling rates. So it is clear that more study is desirable if one is confidently to predict usage in one area from data for another different one.

11.3 Suggested Sample Design for a Before-and-After Study

This section outlines a sampling suggestion for tracking the effects of a pricing change. This suggestion is based on the experience of the authors in conducting the pilot study. It is the design that would have been utilized by the authors had a pricing change occurred in Chicago. We hope that these suggestions may be helpful to researchers in future work.

Master Billing Numbers (MBN's) are the proposed basic sampling units. In the analysis, these units would be equated to households, with the assumption that the observed calling characteristics are attributable to the household with the sampled MBN. We suggest that the residence MBN's be selected and randomly arranged in groups according to the following rotation design, where M represents the month of the rate change, $M - 12$ represents 12 months before the rate change, *etc.* (as explained precisely in Remark (12) below):

Rotation Design

Month $M - 12$	Month $M - 6$	Months $M - 2$ M $M + 1$	Month $M + 6$
A	B	C	D
B	C	D	E
C	D	E	F
D	E	F	G

Remarks

1) Each rotation group might consist of, say, 400 MBN's selected from a current tape of working MBN's. So, for example, group A is a sample of 400 MBN's drawn from a current MBN tape by a systematic random start spaced in such a way that all prefixes are proportionally represented.

2) Group A is observed only in month $M - 12$ and is replaced in month $M - 6$ by Group E, which is selected from the tapes of working MBN's in month $M - 6$. The other rotation groups are selected similarly.

3) Systematic selection bias can enter panel surveys from three sources — differential probabilities of initial selection and of data receipt, attrition, and improper replenishment policies. The first of these sources often has the largest effect. But in this proposed study, the initial selection probabilities are virtually uniform;[2] the probabilities of data receipt are also virtually uniform since telephone measurements will be essentially complete. To minimize the variance of estimates of monthly change, a fixed panel is proposed for the months $M - 2$, M, and $M + 1$. In this period the pricing change is made, and the completely matched sample may give better estimates of change. For this short a period the second source of bias — attrition — is minor: Attrition on the average is usually about 1 percent of the customers per month; in the summer months attrition is somewhat higher, and in winter months somewhat lower. Finally, the period of the sample during which the panel is fixed is short enough (three months) that any deterioration of the sample due to aging (a sample drawn in month $M - 2$ is *not*, strictly speaking, a sample of customers in month $M + 1$) is probably minimal.

4) The rotation scheme permits estimates at the various sampled points of time as well as month-to-month and year-to-year estimates of change.

5) In view of the fairly high correlation through time on some items the optimum rotation for these items is possibly faster than the proposed rotation. But faster rotation would not permit so large a match-up for year-to-year comparisons.

---

[2] Since the sampling units are Master Billing Numbers, a minor problem is that the few households that have more than one line billed separately will be oversampled. So long as the existence of second lines is investigated, this oversampling can be corrected.

6) The real sampling units are Master Billing Numbers and not households. If a customer moves and the MBN is re-issued, this change must be picked up at the processing stage by observation of the change in the customer's name and address.

7) It is always possible that unforeseen events will cause rate-change dates to be moved or canceled. As a result, it may be economically advisable to administer the questionnaires to rotation groups A, B, C, D, E, and F all in month $M - 2$. This means that there will be a lag in the questionnaire data for rotation groups A and B, but it seems desirable to save the questionnaire resources for use at the time at which the rate change is actually made, if at all. An attempt should be made to trace people in rotation groups A and B who have changed their MBN since $M - 12$ or $M - 6$. Let us emphasize that we recommend collecting these questionnaire data two months before the rate change, not at the time of the change; the reason is that the response rate might suffer if some persons are upset by the rate change.

8) The complete set of data to be gathered consists of telephone usage data, billing data, and the demographic questionnaire.

9) At any single point in time the sample design has four active rotation groups of 400 each, or a total of 1600 MBN's. This sample size is larger than the pilot study in order to achieve a larger expected sample size in some relevant demographic subgroups. As in the pilot survey, roughly a 30 percent non-response on the questionnaire can be expected if the survey methodology discussed in the following section is utilized.

10) The total number of observed customers over the four observation periods is seven rotation groups or $7 \times 400 = 2800$ MBN's.

11) This sample has not been designed to study seasonal time trends in telephone usage. It is expensive and time consuming to gather the required data, and the analysis of short-term time trends is generally of secondary interest. But if short-term changes are of interest, additional telephone data could be gathered in months $M - 14$, $M - 11$, and/or $M + 10$, $M + 12$, and $M + 13$.

12) In order to trace the change in customers' calling patterns due to the rate change, one may want a full month's data before the rate change and two months' data after the rate change for every MBN that is in the sample at the time of the rate change. This tracking procedure is complicated, so we shall explain it in terms of an example. Suppose that the rate change begins for all customers on April 1. Because of the use of staggered billing periods, to obtain complete months of calling and billing data that are not *partially* affected by timing, it is necessary

to begin collecting data for some MBN's almost two months before the rate change and to end three months after the rate change. Exhibit 11.1 on the following page portrays by vertical lines the periods for which calling and billing data should be processed. There are two such lines for the MBN's in each of the billing periods, of which there are assumed to be ten. The ten dashed lines on the top portion of the page are for the "before" data, and the ten solid, divided lines on the lower portion are for the "after" data. The latter lines are divided in two to signify that the two months of "after" data for each MBN should be handled separately.

As an example, observe when billing and calling data should be processed for sampled MBN's whose billing period begins on the first of each month. For this group, a complete month of data can be obtained that is unaffected by timing by beginning processing of data on the morning of March 1 and ending on the night of March 31. The "after" data should be processed for April 1 through April 30 and for May 1 through May 31. As another example, for MBN's whose billing period begins on the fourth of each month, "before" data should be processed for February 4 through March 3. (Note that February 4 is the earliest date on which data should be processed.) The "after" data should be processed April 4 through May 3 and May 4 through June 3. (Note that for this group, as well as for all others but one, an entire month separates the end of the "before" data and the beginning of the "after" data.) The rest of the billing period groups are easily read off the table. The last day on which data should be processed is June 27.

Of course, the above dates are given only as an example. If the rate change begins on a day other than April 1, then the entire schedule of data processing should be moved backward or forward by the same number of billing periods as the time of the rate change differs from April 1. Furthermore, if the rate change does not occur on the first day of a month, then the billing period group whose "before" and "after" data are taken back to back should be different from the group whose billing period begins on the first as shown in the diagram. In general terms, "month $M - 2$", corresponding to the "before" data, herein represents the ten billing periods that begin ten through nineteen billing intervals before the rate change; "month M", the first month of "after" data, represents the ten billing periods that begin zero through nine billing intervals after the rate change; and "month $M + 1$", the second month of "after" data, represents the ten billing periods that begin ten through nineteen billing intervals after the rate change. The important points to remember are that the first data to be processed are for the billing period that begins nineteen billing intervals before the

Other Studies and Suggestions for Future Research 361

Exhibit 11.1

When Data Should Be Processed Near Rate Change

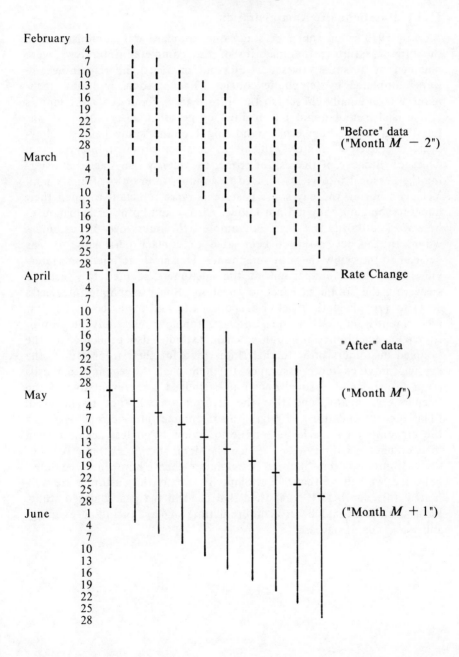

rate change, and the last data to be processed are for the billing period that begins nineteen billing intervals after the rate change.

11.4 Questionnaire Administration and Design

11.4.1 Questionnaire Administration

In 1973 when the pilot study questionnaire was administered to the original sample, the majority of the completed interviews were obtained by personal contact — 365 out of 513. The remaining 148 were completed by telephone contact. The personal interview technique was originally chosen so that information about the respondent's ethnic background might be obtained by personal observation. It was felt that asking ethnic background might offend some people. However, it proved extremely difficult to conduct personal interviews in Chicago. Because of the risk of crime, the agency that was administering the personal interviews refused to conduct interviews after six p.m. Yet even during the day some customers were reluctant to open their doors to someone they did not know. Also, often both heads of household worked during the day. To complete the interviews of customers whom the agency had not been able to contact personally, it was decided to interview them by telephone. During these telephone interviews less than one percent of the customers seemed to object to answering the ethnic background question. Since telephone interviews seem to provide both a better response rate and lower cost, it is our recommendation that the questionnaire for any new study be administered by telephone, although a version should also be mailed to the sampled customers prior to their interviews for their information; the mailing provides more assurance to them that the agency that calls them to ask such personal questions is indeed a representative of the telephone company, and that the information will not be misused. (This recommendation for relying on telephone interviews is based on the presumption that the telephone interview is no less accurate than mail or personal interviews.) Since the demographic characteristics of the customer are very likely to be related to their "home/not-at-home" behavior, we believe that five attempts to contact the customer are warranted: this number is larger than that used in the past Chicago study. Each attempt should be at a different time of day, including evenings and weekends, if necessary.

11.4.2 Questionnaire Design

On the basis of our previous experience, we have designed a new suggested questionnaire. The version that might be mailed to the sampled customers is attached as Appendix 11.A. The same basic design employed in the 1973 Chicago questionnaire is retained. However, questions that confused the respondents have been clarified. Also, questions that did not provide adequate data for analysis have been revised. This section of the chapter explains the reasons for substantial changes in some of the questions from the original 1973 questionnaire.

In the first section of the questionnaire entitled "About Your Service", the number of possible response choices for several of the questions is changed from an even number to an odd number. An odd number of choices allows the respondent more clearly to express a neutral reaction to the particular question. For example, question number one asks the customer to rate the quality of service. In the old questionnaire the choices were: poor, fair, good, excellent; but in the new questionnaire the choices are: very poor, poor, fair, good, very good.

The question on whether the respondent gets his "money's worth" out of his telephone service has been eliminated. The difficulty with the question was a logical one: a person would not continue to subscribe if he did not get his money's worth.

In the second section of the questionnaire, which deals with the use of the telephone, an explanatory sentence has been added to the question that asks the customer how many calls per day he makes to people in the city of Chicago. This sentence now specifies that the customer should exclude incoming calls and calls for which he receives no answer or a busy signal. It was decided to revise this question in this way after it was observed that respondents tend drastically to overestimate the number of calls they make. In localities that do not charge for directory assistance, such calls should also be excluded.

Two questions have been eliminated. These questions asked how the respondent felt about making local and toll calls. The analysis of the 1973 questionnaire suggests that these two questions are insufficiently illuminating to justify their inclusion.

The next section of the questionnaire asks about the respondent's home and family. A question about the type of building in which the respondent lived was eliminated since most of that information is obtained by another question (number 13 in the revised version).

In the old version of the questionnaire, the question that ascertains how many years the respondent has lived in Chicago does not provide us with sufficiently detailed information: 75 percent of the respondents' answers fell into one category — "over 15 years". Now the following choices are provided:

>under 1 year
>1-5 years
>6-10 years
>11-20 years
>21-30 years
>31-40 years
>more than 40 years.

The question on household composition is unchanged, but there is an alternative that might reduce errors: The respondent could be asked to list the exact age and the sex of each family member.

A major change in the questionnaire is that instead of asking questions about the head of the household and the spouse, we request information on the male and female heads of household. The explanation defining the male and female heads of household is as follows: "For example, a husband and wife would be male and female heads; an unmarried woman would be a female head; *etc.*" This change in definition is made to eliminate the displeasure of some respondents to the idea of the household's having only one head. Also, the old questionnaire does not ask all questions about the spouse that are asked about the head (for example, occupation). Now the questionnaire has been revised so that all questions that are asked about the male head are also asked about the female head.

A minor change in wording may be very important. Many respondents to the old questionnaire made no answer to the question on the employment status of the spouse. We speculate that they did not consider it appropriate to label a housewife as "unemployed". Use of the phrase "not employed" may elicit more response.

The above changes in the questionnaire should be useful. They should improve the quality, extent, and ease of analysis of the questionnaire data.

11.4.3 Interviewer Questionnaire

The version of the questionnaire that is suggested to be used by the persons performing interviews is contained in Appendix 11.B.[3] The

---

[3] A separate interviewer's version was not used for the original Chicago sample. Howard Miller, then of IBT Management Sciences, designed one for another study. The version for the present study draws heavily on his work.

interviewer questionnaire differs from that to be mailed to the sampled customers in five respects: Firstly, the front page contains spaces for a respondent's name and telephone number, a form for a record of contact attempts, and a paragraph that the interviewer can use to introduce himself to the respondent. Secondly, many questions are reworded so that each is in a style of a conversational question. Thirdly, some instructions to the interviewer are added. Fourthly, for each question, a "no answer" alternative is included (which is usually coded as zero). Finally, in the right-hand margin, numbers are printed that represent columns of a key punch card. Separate interviewer instructions and coding instructions are contained in Appendix 11.C; these are self-explanatory.[4]

11.5 Suggestions for Future Research

This section of the chapter discusses some suggestions for future research, in addition to what has been suggested in individual chapters. A useful project would be a replication of this study with new data so as to corroborate our findings. Another study would be a straightforward extrapolation of what has already been done — a multivariate regression analysis of time-of-day calling patterns. Besides requiring a larger number of computations, this suggested project needs to be more complex than the regressions in Chapter 6 only to the extent that smoothing of regression coefficients across hours should probably be imposed.

A more ambitious analysis of the Chicago data that would provide important results is the estimation of how the number of business and residence telephones at various distances affects the calling patterns of a household. It should also be possible to estimate the degree to which commonality of each of the various demographic characteristics influences the calling from one household to another. This estimation is possible because there are data on the number of calls that each sampled customer makes to each Central Office Area, and the numbers of people in various demographic groups can be approximated with Census data. The number of business telephones in each Central Office Area can be obtained from telephone company records. The importance of the results would be enhanced if calls to the suburbs and even the toll areas were included in the analysis. The reader should recall that calls from Chicago to the suburbs and to the toll area are charged according to distance and duration whereas calls within Chicago are not; in addition, initiating most calls to the suburbs and to the toll area is more

---
[4] These also draw on similar work by Howard Miller.

expensive than within Chicago. Thus, one may model the volume of calling from a Chicago household to the various Central Office Areas in Chicago, the suburbs, and the toll area as a function of price as well as of distance, of the demographics of the calling household, of the number of business telephones, and of the numbers of people in various demographic groups in each called area. Estimating the model should enable the analyst to predict the volume and geographic patterns of calling from residences in some other city given information for each of its areas on the number of people in each demographic group and the number of business telephones. The analyst could generate a prediction for any rate structure within the range observed for calls from Chicago, including pricing by the number of calls, their duration, and their distance. Notice that out of this analysis could come an estimate of the positive externality that is generated by an additional person's connecting to the telephone network.[5]

A project that could make the above kind of analysis more robust would be the pooling of the Chicago data with similar data from other cities. Hopefully, the rate structure in some of the other cities would be flat rate (*i.e.*, no charge for local calls); and the cities' sizes, density, weather, *etc.*, would vary. It might be possible to estimate from such pooled data how the number of local calls is affected by demographics, by price over a wider range of variation than is observed in Chicago and its suburbs, and by other variables such as the characteristics of the cities themselves.

The authors believe that this book has answered in part the question of how demographic characteristics of households relates to their telephone usage. But there is much room left for further research in this area.

---

[5] A similar approach to estimating the connection externality has been independently suggested by Roger Klein and Robert Willig in some unpublished work at Bell Laboratories.

Appendix 11.A

Suggested Version of the Questionnaire for Mailing

THE TELEPHONE COMPANY
WOULD LIKE YOUR ANSWERS

ABOUT YOUR SERVICE

1. How would you rate the quality of telephone service you are now getting?

    Very poor
    Poor
    Fair
    Good
    Very Good

2. How would you rate the Telephone Company on being concerned about the telephone service problems of the individual customers?

    Very poor
    Poor
    Fair
    Good
    Very Good

3. How would you rate the Telephone Company on making it easy and pleasant to do business with the company?

    Very poor
    Poor
    Fair
    Good
    Very Good

4. Do you feel that the cost for telephone service has gone up much more, more, about the same, less, or much less than the cost of other goods and services in the past few years?

    Much more

    More

    About the same

    Less

    Much less

5. Do you feel you are generally billed accurately for the number of message units used?

    Yes, billed accurately

    No, billed for **more** than used

    No, billed for **less** than used

ABOUT YOUR TELEPHONE AND HOW YOU USE IT

6. How long have you had your present telephone number?

    Under six months

    Six months to 1 year

    1 to 2 years

    More than 2 years

7. On an average day, about how many calls do you and others living here make to people **within** the city of Chicago? (Please **exclude** calls that others make to you, or calls that you attempt but which end in a busy signal or unanswered.)

    _____ calls
    (your best estimate)

8. Of **all** the telephone calls you and others living here make (including long distance), about what percent of these calls are to people living within the city of Chicago?

25% or less

26% to 50%

51% to 75%

76% to 90%

91% to 100%

9. On calls made by you and by others living here to people in the city of Chicago, would you say that the average length of these calls is:

Under 5 minutes

5 to 10 minutes

Over 10 minutes

10. How many telephone lines do you have in your home? Each line has a different telephone number; for example _____ is one line and 555-3398 would be another. (Most customers have one line.)

_____
(Please specify number)

11. How many phones do you have in your home?

_____
(Please specify number)

12. Do you have "Touch-Tone"® (pushbuttons) in place of a dial on your telephone?

Yes

No

ABOUT YOUR FAMILY AND HOME

13. Do you or other members of your household own or rent your present residence?

Own	*Rent*
House	Apartment
Cooperative	House
Condominium	Room in a house
Multi-Family	Room in an apartment
Mobile or trailer home	Mobile or trailer home
	Other

14. How long have you lived at this address?

- Under 6 months
- Six months to 1 year
- 1 to 2 years
- 3 to 5 years
- 6 to 10 years
- 11 to 15 years
- More than 15 years

15. How long have you lived in the city of Chicago?

Under 1 year

1 to 5 years

6 to 10 years

11 to 20 years

21 to 30 years

31 to 40 years

More than 40 years

16. How many times have you moved in the past five years?

None

Once

Twice

Three times

Four or more times

17. Do you (or any members of your household) own or rent a second home or other living quarters that you occupy during part of the year?

Yes

No

18. Are you a head of the household? (For example, a husband and wife would be the male head and female head; an unmarried woman would be the female head, *etc.*)

Yes

No

19. What is the marital status of the head or heads of the household?

    Single

    Married

    Separated

    Divorced

    Widowed

20. What is the age of the head or heads of the household?

Male Head, if any	Female Head, if any
Under 25	Under 25
25 to 34	25 to 34
35 to 44	35 to 44
45 to 54	45 to 54
55 to 64	55 to 64
65 and over	65 and over

21. If you are not a head of household, are you male or female?

    Male

    Female

22. If you are not a head of household, what is your age?

    Under 25

    25 to 34

    35 to 44

    45 to 54

    55 to 64

    65 and over

23. Please fill in the table below for **all** family members and others who normally live in your home at least 2 months a year, including yourself. For each age group, write the number of males and females normally living at home.

Age Group	Number of Males	Number of Females
Under 10	_____	_____
10 to 12	_____	_____
13 to 15	_____	_____
16 to 18	_____	_____
19 to 24	_____	_____
25 to 34	_____	_____
35 to 54	_____	_____
55 to 64	_____	_____
65 and over	_____	_____

24. How many people live here during **each** of the seasons of the year?

	Number
Summer	_____
Fall	_____
Winter	_____
Spring	_____

25. What is the highest grade attended or degree received by the head or heads of the household?

Male Head, if any	Female Head, if any
Eighth grade or less	Eighth grade or less
Some high school	Some high school
High school graduate	High school graduate
Some college	Some college
College graduate	College graduate
Some graduate school	Some graduate school
Graduate degree	Graduate degree

26. What is the present employment status of the head or heads of the household?

Male Head, if any	Female Head, if any
Employed full time	Employed full time
Employed part time	Employed part time
Not employed, looking for work	Not employed, looking for work
Not employed, not looking for work	Not employed, not looking for work
Retired	Retired

27. If presently "employed" do the head or heads of the household use the home as a place of business?

Employed Male Head, if any	*Employed Female Head, if any*
Yes, all the time	Yes, all the time
Yes, occasionally	Yes, occasionally
No	No

28. If presently employed, in what type of business, industry, or service is the company or employer of the head or heads presently engaged? Please be specific, *e.g.*, steel manufacturing, education, textile wholesaling, state government, hardware retailing, *etc*. If the head or heads of the household are self-employed, please also say so.

Employed Male Head, if any

_____

_____

Employed Female Head, if any

_____

_____

29. If presently employed, what is the title or position of the head or heads of the household? (*E.g.*, foreman, doctor, carpenter, teacher, salesman, machinist, *etc.*)

Employed Male Head, if any

_____

_____

Employed Female Head, if any

_____

_____

30. What is your ethnic background?

    Black

    Oriental

    Spanish

    White

    Other

31. Please check off your combined approximate household income for **19xx** (that is, total wages, interest, dividends, *etc.*, of all persons presently in your household before taxes and other deductions).

    Under $5,000

    $5,000 to $8,999

    $9,000 to $14,999

    $15,000 to $19,999

    $20,000 to $29,999

    $30,000 or more

Appendix 11.B

Suggested Version of the Questionnaire for Interviewer

LOCAL USAGE STUDY

Name _____

Telephone Number _____

1-
2-
3-
4-
5-
6-
7-

ATTEMPT RECORD

Call	Date	Time	Interviewer No.	Completed Interview	No answer	Busy	Refusal	Desig. Resp. Not Avail.	Non-working No.	Other (specify)
1										
2										
3										
4										
5										

Hello. Is this the _____ residence? I'm _____ of _____. We are calling for the Illinois Bell Telephone Company about the letter and questionnaire that you recently received in the mail. I would like to obtain opinions concerning these questions. Do you have the questionnaire we sent you? (IF "NO", ATTEMPT TO COMPLETE THE INTERVIEW ANYWAY.)

First, I'd like to ask about your service

1. How would you rate the quality of telephone service you are now getting? Would you say that it is very poor, poor, fair, good, or very good?

NO ANSWER	0	8-
Very poor	1	
Poor	2	
Fair	3	
Good	4	
Very good	5	

2. How would you rate the Telephone Company on being concerned about the telephone service problems of the individual customers? Would you say ... (READ LIST)

NO ANSWER	9	9-
Very poor	1	
Poor	2	
Fair	3	
Good	4	
Very good	5	

3. How would you rate the Telephone Company on making it easy and pleasant to do business with the company? Would you say ... (READ LIST)

NO ANSWER	0	10-
Very poor	1	
Poor	2	
Fair	3	
Good	4	
Very good	5	

4. Do you feel that the cost for telephone service has gone up much more, more, about the same, less, or much less than the cost of other goods and services in the past few years?

NO ANSWER	0	11-
Much more	1	
More	2	
About the same	3	
Less	4	
Much less	5	

5. Do you feel you are generally billed accurately for the number of message units used? (IF NO:) Do you feel you are billed for more than the calls you make or for less than the calls you make?

NO ANSWER	0	12-
Yes, billed accurately	1	
No, billed for **more** than used	2	
No, billed for **less** than used	3	

Now about your telephone and how you use it.

6. How long have you had your present telephone number? Would you say ... (READ LIST)

NO ANSWER	0	13-
Under six months	1	
Six months to 1 year	2	
1 to 2 years	3	
More than 2 years	4	

7. On an average day, about how many calls do you and others living here make to people **within** the city of Chicago? (Please **exclude** calls that others make to you, or calls that you attempt but which end in a busy signal or unanswered.)

NO ANSWER	0	14-
_____		
(NUMBER)		

8. Of **all** the telephone calls you and others living here make (including long distance), about what percentage of these calls are to people living within the city of Chicago? Would you say ... (READ LIST)

NO ANSWER	0	15-
25% or less	1	
26% to 50%	2	
51% to 75%	3	
76% to 90%	4	
91% to 100%	5	

9. On calls made by you and by others living here to people in the city of Chicago, would you say that the average length of these calls is ... (READ LIST)

NO ANSWER	0	16-
Under 5 minutes	1	
5 to 10 minutes	2	
Over 10 minutes	3	

10. How many telephone lines do you have in your home? Each line has a different telephone number; for example _____ is one line and 555-3398 would be another. (Most customers have one line.)

NO ANSWER	0	17-
_____		
(NUMBER)		

11. How many phones do you have in your home?

NO ANSWER	0	18-
_____		
(NUMBER)		

12. Do you have "Touch-Tone" (pushbuttons) in place of a dial on your telephone?

NO ANSWER	0	19-
Yes	1	
No	2	

Now about your family and home.

13. Do you or other members of your household own or rent your present residence?

    (IF OWN:)   **(IF RENT:)**

    Is that a ... (READ LIST)   Is that an ... (READ LIST)

OWNS, BUT NO ANSWER ON TYPE	0	RENTS, BUT NO ANSWER ON TYPE	0	20-
House	1	Apartment	1	21-
Cooperative	2	House	2	
Condominium	3	Room in a house	3	
Multi-family dwelling	4	Room in an apartment	4	
Mobile or trailer home	5	Mobile or trailer home	5	
RENTS	6	Other	6	
ANSWERS NEITHER OWN NOR RENT	7	OWNS	7	
		ANSWERS NEITHER OWN NOR RENT	8	

14. How long have you lived at this address? Would you say ... (READ LIST)

NO ANSWER	0	22-
Under 6 months	1	
Six months to 1 year	2	
1 to 2 years	3	
3 to 5 years	4	
6 to 10 years	5	
11 to 15 years	6	
More than 15 years	7	

15. How long have you lived in the city of Chicago? Would you say ... (READ LIST)

NO ANSWER	0	23-
Under 1 year	1	
1 to 5 years	2	
6 to 10 years	3	
11 to 20 years	4	
21 to 30 years	5	
31 to 40 years	6	
More than 40 years	7	

16. How many times have you moved in the past five years? Would you say ... (READ LIST)

NO ANSWER	0	24-
None	1	
Once	2	
Twice	3	
Three times	4	
Four or more times	5	

17. Do you (or any members of your household) own or rent a second home or other living quarters that you occupy during part of the year?

NO ANSWER	0	25-
Yes	1	
No	2	

18. Are you a head of the household? (For example, a husband and wife would be the male head and female head; an unmarried woman would be a female head; and so forth.)

NO ANSWER	0	26-
Yes	1	
No	2	

19. What is ⎰ your marital status?   NO ANSWER   0   27-
         ⎱ the marital status of the head or
           heads of the household?   Single   1
   (READ LIST)
                                                  Married   2

                                                  Separated 3

                                                  Divorced   4

                                                  Widowed   5

IF "MARRIED", THEN ASK ALL QUESTIONS ABOUT THE HEAD FOR BOTH THE MALE HEAD AND THE FEMALE HEAD.

20. ⎧ What is your age?
    ⎨ What is your age and the age of your ⎰ husband?
    ⎩   ⎱ wife?
      What is (are) the age(s) of the head(s) of the household?
      (READ LIST)

	MALE HEAD, IF ANY	FEMALE HEAD, IF ANY	
NO ANSWER	0	0	28-
Under 25	1	1	29-
25 to 34	2	2	
35 to 44	3	3	
45 to 54	4	4	
55 to 64	5	5	
65 and over	6	6	

IF RESPONDENT IS A HEAD OF HOUSEHOLD, SKIP TO QUESTION 23.

21. (IF RESPONDENT IS NOT   RESPONDENT IS HEAD,   0   30-
    HEAD OF HOUSEHOLD)   OR NO ANSWER
    Are you male or female?   Male   1

                                                Female 2

22. **(IF RESPONDENT IS NOT HEAD OF HOUSEHOLD)**
What is your age?
(READ LIST)

RESPONDENT IS HEAD, OR NO ANSWER	0
Under 25	1
25 to 34	2
35 to 44	3
45 to 54	4
55 to 64	5
65 and over	6

31-

23. Thinking about **all** family members, including yourself and others who normally live in your home at least 2 months a year, please tell me how many males and females for each of the following age groups are normally living at home. (READ LIST)

	MALES (NUMBERS)		FEMALES (NUMBERS)	
Under 10	_____	32-	_____	41-
10 to 12	_____	33-	_____	42-
13 to 15	_____	34-	_____	43-
16 to 18	_____	35-	_____	44-
19 to 24	_____	36-	_____	45-
25 to 34	_____	37-	_____	46-
35 to 54	_____	38-	_____	47-
55 to 64	_____	39-	_____	48-
65 and over	_____	40-	_____	49-

INCOMPLETE ANSWER [ ]

(IF THERE ARE NONE IN AN AGE GROUP, LEAVE BLANK FOR THAT GROUP).

24. How many people live here during **each** of the seasons of the year? How many persons live in your home during the summer season? During the fall season? The winter season? And the spring season?

Summer	_____ (NUMBER)	50-
Fall	_____ (NUMBER)	51-
Winter	_____ (NUMBER)	52-
Spring	_____ (NUMBER)	53-

25. What is the highest grade attended or degree received by

{ you?
{ you and your { husband?
{ { wife?
{ the head(s) of the household?
(READ LIST)

	MALE HEAD, IF ANY	FEMALE HEAD, IF ANY	
NO ANSWER	0	0	54-
Eighth grade or less	1	1	55-
Some high school	2	2	
High school graduate	3	3	
Some college	4	4	
College graduate	5	5	
Some graduate school	6	6	
Graduate school	7	7	

26. ⎧ What is your present employment status?
   ⎨ What is the present employment status of you and your ⎰ husband?
   ⎩   ⎱ wife?
     What is the present employment status of the head(s) of the household?
   (READ LIST)

	MALE HEAD, IF ANY	FEMALE HEAD, IF ANY	
NO ANSWER	0	0	56-
Employed full time	1	1	57-
Employed part time	2	2	
Not employed, looking for work	3	3	
Not employed, not looking for work	4	4	
Retired	5	5	

27. **(IF MALE HEAD IS PRESENTLY EMPLOYED)**

⎧ Do you   ⎫
⎨ Does your husband   ⎬ use the home as a place of business?
⎩ Does the male head of the household   ⎭

(IF YES:) All the time or occasionally?

(IF FEMALE HEAD IS PRESENTLY EMPLOYED)

$\left\{\begin{array}{l}\text{Do you}\\ \text{Does your wife}\\ \text{Does the female head of the household}\end{array}\right\}$ use the home as a place of business?

(IF YES:) All the time or occasionally?

	EMPLOYED MALE HEAD, IF ANY	EMPLOYED FEMALE HEAD, IF ANY	
NO ANSWER	0	0	58-
Yes, all the time	1	1	59-
Yes, occasionally	2	2	
No	3	3	
NOT EMPLOYED	4	4	

28. **(IF MALE HEAD IS PRESENTLY EMPLOYED)**

In what type of business, industry, or service is
$\left\{\begin{array}{l}\text{your}\\ \text{your husband's}\\ \text{the household's male head's}\end{array}\right\}$ company or employer engaged?

Please be specific; for example, steel manufacturing, education, textile wholesaling, state government, hardware, retailing, and so forth.

$\left\{\begin{array}{l}\text{Are you}\\ \text{Is your husband}\\ \text{Is the male head of household}\end{array}\right\}$ self-employed?

EMPLOYED MALE HEAD, IF ANY

_____ 60-
_____ 61-
_____ 62-
_____ 63-

(IF FEMALE HEAD IS PRESENTLY EMPLOYED)

In what type of business, industry, or service is
$\left\{\begin{array}{l}\text{your}\\ \text{your wife's}\\ \text{the household's female head's}\end{array}\right\}$ company or employer engaged?

Please be specific; for example, steel manufacturing, education, textile wholesaling, state government, hardware retailing, and so forth.

$\left\{\begin{array}{l}\text{Are you}\\ \text{Is your wife}\\ \text{Is the female head of household}\end{array}\right\}$ self-employed?

EMPLOYED FEMALE HEAD, IF ANY

_____ 64-
_____ 65-
_____ 66-
_____ 67-

29. **(IF MALE HEAD IS PRESENTLY EMPLOYED)**

What is $\left\{\begin{array}{l}\text{your}\\ \text{your husband's}\\ \text{the household's male head's}\end{array}\right\}$ title or position?

(For example, foreman, doctor, carpenter, teacher, salesman, machinist, and so forth.)

EMPLOYED MALE HEAD, IF ANY

_____ 68-
_____ 69-
_____ 70-

(IF FEMALE HEAD IS PRESENTLY EMPLOYED)

What is $\left\{\begin{array}{l}\text{your}\\ \text{your wife's}\\ \text{the household's female head's}\end{array}\right\}$ title or position?

(For example, foreman, doctor, carpenter, teacher, salesman, machinist, and so forth.)

EMPLOYED FEMALE HEAD, IF ANY

_____ 71-
                                                72-
_____ 73-

30. What is your ethnic background? Are you ... (READ LIST)

NO ANSWER	0	74-
Black	1	
Oriental	2	
Spanish Speaking	3	
White	4	
Other	5	

31. Which of the following categories represents your combined approximate household income for 19xx (that is, total wages, interest, dividends, *etc.*, of all persons presently in your household before taxes and other deductions). Was it ... (READ LIST)

NO ANSWER	0	75-
Under $5,000	1	
$5,000-$8,999	2	
$9,000-$14,999	3	
$15,000-$19,999	4	
$20,000-$29,999	5	
$30,000 or more	6	

Thank you very much for your answers, Mr. (Miss, Mrs., Ms.) _____.
I'm sure the Telephone Company appreciates your taking your time to help us.

INTERVIEWER CODE   76-

Appendix 11.C

Interviewer Instructions

PURPOSE

The purpose of this study is to determine how telephone calling patterns are related to the characteristics of households.

PRELIMINARY INFORMATION

Prior to the telephone interview, each respondent will receive by mail a copy of the questionnaire to look over. They will not be requested to fill out the questionnaire but to keep it handy for the telephone interview. A copy of the version of the questionnaire that is mailed to them is attached for your reference.

QUALIFICATION OF RESPONDENT

Make sure that you get either the male or female head of the household or a responsible adult (over 18 years old) familiar with the household's telephone usage.

YOUR QUESTIONNAIRE

Most of your instructions appear within your version of the questionnaire in capital letters. Do not read to the respondent these instructions or the "NO ANSWER" alternatives!

For most responses, please circle the number for the appropriate response; but for questions 7, 10, 11, 23, 24, 28, and 29, write in the response where specified.

Do not force the respondent or badger him or her in any way to be interviewed. Be polite. The questionnaire deals with information that is considered sensitive and private by some people. Therefore, tact must be used. Accept a refusal to any question politely. It is desired to complete as many interviews as possible. Therefore, if a respondent refuses on the basis that he or she considers the demographic information to be too personal, an attempt should at least be made to complete the portion of the questionnaire dealing with telephone usage.

QUESTIONS ABOUT SERVICE AND USE OF TELEPHONE

For the most part, these questions are straightforward and self-explanatory. If a respondent does not understand a question, reread the question. Do not lead the respondent or attempt to explain the question in your own words. If the respondent does "not know" a response, record as "NO ANSWER".

Question 7:

Only write in a "0" for the number of calls if a respondent does not answer. If his answer is less than one call per day — even none — then write in one.

Question 8:

If a respondent does not understand the question, then repeat, "About what percentage of all your calls, including long distance, are local calls?"

Question 10:

Insert the respondent's telephone number in the "blank" as an example of a telephone line. It is suggested that the telephone number be listed prior to beginning the interview.

QUESTIONS ABOUT FAMILY AND HOME

Questions 19, 20, and 25 through 29 deal with the head or heads of household. For the purpose of this questionnaire, note that the typical household, with a husband and wife, has *both* a male head *and* a female head. (In this case, make sure that all questions that affect both male and female heads have answers for both.) Any other type of household would ordinarily have only one head.

For the questions dealing with the heads, the phrasing should vary depending on the situation. If the respondent is the only head, then ask the question in the second person singular (*e.g.*, "What is your ..."); if the respondent is one of two heads, then ask about both (*e.g.*, "What is your and your husband's ..." or "What is your and your wife's ..."); if the respondent is not a head, then ask in the third person (*e.g.*, "What is the head's ..." or "What are the heads' ...").

Question 23:

Record an answer for each category for each sex. Some households have boarders. These boarders should be excluded from the tables for this question and for number 24 unless they are permitted to make calls from the household's telephone.

Question 26:

If neither head is employed full time or part time, then SKIP TO QUESTION 30.

If one or both are employed be sure to ask questions 28 and 29 for each *employed* head. If a person takes care of his or her own house or children, *etc.*, and does not earn an outside income, then either "not employed" or "retired" may be an appropriate entry.

Question 28:

This relates to the occupations of the heads of the household. You must be specific as to the type of business, industry, or service where the person is employed.

For example: If you are interviewing the female head of the household, and she says that her husband "is an accountant", this is his position within the company and would be recorded as the answer to Question 29. Then ask "What *type* of company is he an accountant for?" (*i.e.*, insurance company, car manufacturer, *etc.*) and record this information as an answer for Question 28.

If the male or female head of household has more than one job, then record both and indicate which is his or her primary job.

If the respondent previously mentioned "retired" or "not employed" in Question 26, then write in that status where appropriate in Questions 28 and 29 for the member who is retired or not employed.

If the male or female head of household is a student or on welfare, then write it in — these are complete answers within themselves.

COMPLAINTS

If a respondent complains about the questionnaire or about his or her telephone service, please record the complaint, attach it to the questionnaire, and tell the respondent that you will pass on the complaint to the Telephone Company.

PRIVACY

If a respondent seems to be hesitant about answering any of the questions because it is invading his privacy, reassure him that the results of the questionnaire are to be kept strictly confidential. Such an assurance could, for example, be made by reading to him the paragraph on that subject in the letter which was mailed to him with the questionnaire.

Coding Instructions

The numbers printed in the right margin represent the columns of a punch card. Make a single-digit entry for *every* one of these numbers. For most of the questions, write in the margin the number which is circled as an answer. The exceptions are as follows:

The telephone number should be coded as the first seven entries.

Questions 8, 11, 12:

Code the number written in unless it is greater than 9, in which case code 9.

Questions 14, 21, 26, 27, and 28:

Each of these is two questions in one. The part to the left should be coded next to the first number in the right margin; the part to the right should be coded second. For example, in question 14, an answer to the "IF OWN" part of the questions should be coded next to "22-".

Questions 24 and 25:

Code blanks as zeroes. Code anything larger than 9 as 9.

Question 29:

This question, along with question 30, necessitates the most care. Look up each business in the attached list labeled "Standard Industrial Classifications", and write down the four digits corresponding to it. The first two digits are the most important.

Question 30:

Look up each position in the attached list labeled "Occupational Classification System", and write down the three digits corresponding to it. Again, the first two digits are the most important.

The final step is to record the digit which represents the interviewer, utilizing the list provided by your agency.

Bibliography

Takeshi Amemiya, "The Estimation of the Variances in a Variance-Components Model", *International Economic Review,* v. 12, no. 1 (February, 1971), pp. 1-13.

Takeshi Amemiya, "Regression Analysis When the Variance of the Dependent Variable is Proportional to the Square of Its Expectation", *Journal of the American Statistical Association,* v. 68, no. 344 (December, 1973), pp. 928-34.

Pietro Balestra and Marc Nerlove, "Pooling Cross Section and Time Series Data in the Estimation of a Dynamic Model: The Demand For Natural Gas", *Econometrica,* v. 34, no. 3 (July, 1966), pp. 585-612.

Ralph Braid, "An Evaluation of Estimation Techniques for Logit Probability Models When Applied to Aggregate Data on Choice Frequency", unpublished paper, Massachusetts Institute of Technology (November, 1978).

D. A. S. Fraser, *Statistics: An Introduction* (New York: John Wiley and Sons, Inc., 1960), pp. 173-84.

Jerry A. Hausman, "Specification Tests in Econometrics", *Econometrica,* v. 46, no. 6 (November, 1978), pp. 1251-72.

Charles R. Henderson, "Estimation of Variance and Covariance Components", *Biometrics,* v. 9, no. 2 (June, 1953), pp. 226-52.

Honeywell Information Systems, Inc., "DSS190 and DSS190B Disc Storage Subsystem Reference Manual, Series 6000", Publication DB37.

Honeywell Information Systems, Inc., "Integrated Data Store", Publication BR69.

International Business Machines Corp., "Information Management System/360, Version 2", Publication SH20-0910/11/12/14/15.

Norman L. Johnson and Samuel Kotz, *Distributions in Statistics: Continuous Univariate Distributions — 1* (Boston: Houghton Mifflin, 1970).

Journal of Public Economics, v. 6, nos. 1 and 2 (July-August, 1976).

Leo Katz, "Planning Is for People", *IEEE Spectrum,* v. 12, no. 2 (February, 1975), pp. 57-60.

B. W. Lindgren and G. W. McElrath, *Introduction to Probability and Statistics* (New York: Macmillan, 1966).

G. S. Maddala and T. D. Mount, "A Comparative Study of Alternative Estimators for Variance Components Models Used in Econometric Applications", *Journal of the American Statistical Association,* v. 68, no. 342 (June, 1973), pp. 324-28.

Robert McGill, John W. Tukey, and Wayne A. Larsen, "Variations of Box Plots", *The American Statistician,* v. 32, no. 1 (February, 1978), pp. 12-16.

Yair Mundlak, "On the Pooling of Time Series and Cross Section Data", *Econometrica,* v. 40, no. 1 (January, 1978), pp. 69-85.

Marc Nerlove, "Further Evidence on the Estimation of Dynamic Economic Relations from a Time Series of Cross Sections", *Econometrica,* v. 39, no. 2 (March, 1971), pp. 359-82.

Bruce M. Owen and Ronald Braeutigam, *The Regulation Game* (Cambridge, Mass.: Ballinger, 1978).

Pacific Telephone and Telegraph Co., *Lifeline 1976: Characteristics of Residence Subscribers,* General Administration Accounting, Market Research and Statistics, Project 6-12 (June, 1976).

Lewis J. Perl, "Economic and Demographic Determinants of Residential Demand for Basic Telephone Service", National Economic Research Associates, Inc. (March 28, 1978).

C. Radhakrishna Rao, "Estimating Variance and Covariance Components in Linear Models", *Journal of the American Statistical Association,* v. 67, no. 337 (March, 1972), pp. 112-15.

Shayle R. Searle, *Linear Models* (New York: John Wiley and Sons, 1971).

George J. Stigler, "The Theory of Economic Regulation", *Bell Journal of Economics and Management Science,* v. 2, no. 1 (Spring, 1971), pp. 3-21.

William E. Taylor, "Pooling Time Series and Cross Section Data: Exact Finite Sample Results", *Journal of Econometrics,* forthcoming.

James Tobin, "Estimation of Relationships for Limited Dependent Variables", *Econometrica,* v. 26, no. 1 (January, 1958), pp. 24-36.

John W. Tukey, *Exploratory Data Analysis* (Reading, Mass.: Addison-Wesley, 1977).

Thomas D. Wallace and Ashiq Hussain, "The Use of Error Components in Combining Cross Section With Time Series Data", *Econometrica,* v. 37, no. 1 (January, 1969), pp. 55-72.

Martin B. Wilk and Ramanathan Gnanadesikan, "Probability Plotting Methods for the Analysis of Data", *Biometrika,* v. 55, no. 1 (March, 1968), pp. 1-17.

Edward E. Zajac, *Fairness or Efficiency: An Introduction to Public Utility Pricing* (Cambridge, Mass.: Ballinger, 1978).

Index

age of the head of household, 6-8, 82, 85, 86, 88, 89, 91, 167, 169, 171-173, 175, 177, 179, 180, 227-232, 280, 288, 290, 306, 309-312, 314, 317
Amemiya, Takeshi, 145
American Telephone and Telegraph Co., xvi, xvii, 30
Ancmon, Elsa M., xvii, 75, 133, 165, 219

Balestra, Pietro, 145
basic service charge, 276
    defined, 43
Becker, Richard A., xvii, 133
Bell Telephone Laboratories, xvi, xvii, 67, 83, 356, 366
bias, 13, 26, 44, 46, 134, 158, 287, 307
    rounding, 17, 82-85, 136, 169, 311
bill, computed monthly, 276, 278-280, 286-288
billing data, 40, 41, 43-45, 58, 59, 276, 277, 359, 361
    periods, 278
box plot, defined, 78, 80, 306, 307
Braeutigam, Ronald, 12
Braid, Ralph, 145

Brandon, Belinda B., xvii, 1, 29, 75, 133, 150, 165, 219, 275, 355
Brandon, Paul S., xvii, 1, 75, 133, 165, 275, 355

call, defined, 41
    local, defined, 41
    suburban, defined, 41, 87
    toll, defined, 41
calling data, 41, 43-46, 58-60, 66, 70, 359, 361
causality, 9, 220
Census, 22, 24, 25, 365
Central Office Areas, 19, 22-28, 37, 221, 224, 225, 278, 280, 288, 290
charges, total, 8, 38
    toll, 36, 38, 45
Chicago Association of Commerce and Industry, 24, 25
children, 175, 176
Cincinnati Bell Telephone Co., 355, 357
Cohen, Aya, 83
Collins, S. M., 356
community of interest, 76, 233, 315, 318, 365
computed monthly bill, 276, 278-280, 286-288

concessions, 36, 43-45, 304
conversation time, total local, 5-7, 36, 38, 45, 85, 86, 152, 157, 171, 172, 177, 178
    local and suburban, 8, 225, 228-230
    suburban, 5, 6, 89, 172, 178
    toll, 9, 309, 310
customer-estimated local usage, 7, 86

data base, 57-74
data, billing, 40, 41, 43-45, 58, 59, 276, 277, 359, 361
    calling, 41, 43-46, 58-60, 66, 70, 359, 361
    toll, 43-45
demography of Chicago, 21, 23-25, 27, 28
Devlin, Susan J., 70, 83, 303
distance, 8, 219-233
    average, 8
    of local and suburban calls, 231, 232
    defined, 222
duration of local and suburban calls, 8, 225, 228, 230
    local calls, 5-7, 38, 45, 82-85, 87, 148, 152, 156, 169, 170, 176, 177, 314, 316
    suburban calls, 5, 88, 89, 171, 178, 315, 316, 317
    toll calls, 310, 311

economic efficiency, 11
education, 140, 278
employment status, 140, 156, 157, 278
equivalent days, 135

error checking, 38, 40, 41

Fagal, Frederick, 2
Federal Communications Commission, 356
Finck, Carole C., 75
flat rate, 2
Fowlkes, Edward B., xvii, 275
Fraser, D. A. S., 279
future research suggestions, 136, 144, 158, 180, 181, 233, 318, 319, 357-366

Garfinkel, Lawrence, xvii
Gnanadesikan, Ramanathan, xvii, 80, 279
Goodman, Michael L., 19
Greenwald, Bruce C. N., xvii, 133
Groff, Robert H., 75, 275
Gross, Alan M., xvii

Hausman, Jerry A., xvii, 133
Helmert orthogonal transformation, 279
Henderson, Charles R., 145
Hevrdejs, R. J., 33
Hoeppner, Edmund C., xvii
Honeywell Information Systems, 63, 67
household composition, 6, 8, 140, 156, 157
    size, 8, 280, 288, 290
households, number of, 24
housing values, 25
Hussain, Ashiq, 145-147

Illinois Bell Telephone Co., xv, xvi, xvii, 22, 24, 25, 31, 35, 40, 41, 44, 364
Illinois Commerce Commission, xv

impact of pricing changes, 9, 11, 14
income, 6-10, 82, 85, 86, 88-90, 156, 157, 169, 170, 172, 175-177, 179, 227-233, 287, 290, 305, 308, 309, 311, 312, 314-316
    elasticity, 76
    redistribution, 11, 12
inferences beyond the sample, 26, 134, 154, 155, 158
Infosino, William J., xvii, 133, 136, 356
interactions, 82, 139, 140, 278, 280, 288, 290
International Business Machines, 63

Johnson, Norman L., 285

Katz, Leo, 2
Kennedy, James N., xvii
Klein, Roger W., xvii, 133, 366
Koenker, Roger W., xvii, 133
Kotz, Samuel, 285

Larsen, Wayne A., xvii, 77, 78, 80, 275
Lindgren, B. W., 93
Little, Robert, 2
local call, defined, 41

Maddala, G. S., 145
Mallows, Colin L., xvii, 133
marital status, 8, 280, 290
Market Research Information System, 30
Martin, Patti, xvii
McElrath, G. W., 93
McGill, Robert, xvii, 57, 77, 78, 80

message units, defined, 4
    suburban, 6, 7, 38, 45, 89, 90, 152, 157, 172, 178, 179
    total, 6, 36, 38, 45, 90-92, 152, 157, 173, 180, 225, 229, 230
message-unit allowance, 91, 92
methodology, 14-18
Miller, Howard, 364, 365
minute-miles, average, of local and suburban calls, 231, 233
    total, of local and suburban calls, 231, 232
Mitchell, Bridger, 356
Mount, T. D., 145
multiple comparisons problem, 15, 16
Mundlak, Yair, 145

National Economic Research Associates, 2
National Science Foundation, 356
Nerlove, Marc, 145
New York Telephone Co., xv, 2
Normal distribution, 279
number of households, 24
number of local and suburban calls, 224, 227, 230
    local calls, 5-7, 36, 38, 45, 81, 82, 86, 87, 114, 118, 119, 150-152, 156, 159, 162-164, 169, 174-176, 314-317, 356, 357
    suburban calls, 5, 6, 88, 171, 178, 314-317
    toll calls, 9, 306-309

Opinionmetrics, 33
Owen, Bruce M., 12

Pacific Telephone and Telegraph Co., 12, 355, 357
Patterson, I. Lester, 70, 303
Pavarini, Carl, xvii, 133, 136
Perl, Lewis J., 2
Pollak, Henry O., xvii
position of the head of household, 156, 157
power transformation, 279
prediction, 13, 26
proximity, 224, 225
    defined, 222

questionnaire, 30-41, 43, 46-55, 58-60, 157, 362-365, 367-393

race, 5-9, 24, 25, 34, 81, 85, 88-90, 156, 157, 169-173, 176, 177, 179, 227-233, 280, 287, 288, 290, 305-308, 310-315
Rand Corporation, 356
random access, 62, 63, 67, 73, 74
Rao, C. Radhakrishna, 145
rate periods, 169, 170, 172, 173
    defined, 167
    structure in Chicago, 4, 91, 134
redistributive effect, 11
refusal rates, 38, 39
rents, 25
representativeness of the sample, 223, 224
repricing, 9, 11
response rate, questionnaire, 34-39
revenue effect, 11

sample of customers, 21, 357-359
Scatter Storage System, 68-70
Searle, Shayle R., 145
Shugard, Mary H., xvii, 133, 136

Sinden, Frank W., xvii
Southern New England Telephone Co., 2, 355, 356
Stafford, Richard G., xvii, 133, 136
Stigler, George J., 1, 12
Subscriber Attitude Measurement, 30
suburban call, defined, 41, 87
suburban/no-suburban, 314
switching equipment, 19, 20, 23-25, 27, 28
Szabo, Stephanie, xvii, 133

Taylor, William E., xvii, 133, 146, 275
Technology & Economics, 356
Terpenning, Irma, xvii, 133
time of day, 7, 165, 173-180, 231-233
Tobin, James, 282
Tobit analysis, 282-285
toll, 9, 303-319
    call, defined, 41
    charges, 9, 36, 38, 45, 311, 312
    data, 43-45
toll/no-toll, 9, 305, 306
transformation, Helmert orthogonal, 279
    power, 279
Tukey, John W., 77, 78, 80
Tukey, Paul A., xvii

usage sensitive pricing, 3

vertical service charge, 8, 38, 276, 278, 281-285, 288-290
    defined, 43

Wachter, Kenneth, xvii, 180

Wallace, Thomas D., 145-147
Warner, Jack L., 67
Wilcoxon-Mann-Whitney rank sum test, 93
Wilk, Martin B., 80
Williams, Wm. H., 19, 355
Willig, Robert, 366
years lived in Chicago, 8, 9, 140, 280, 290, 305, 306, 309-312, 314, 316, 317
Zajac, Edward E., xvii, 11

About the Editor

Belinda B. Brandon is a member of technical staff at Bell Laboratories, Murray Hill, N.J., in the Economics Research Center. She was born in Seattle, Washington in 1947. She received the B.A. degree from the University of Washington in 1968 and the Ph.D. degree in economics from the Massachusetts Institute of Technology in 1972. At that time, she joined Bell Laboratories. While there, she also taught courses in the economics department of New York University. She has published articles on corporate mergers. Her current interests include demand analysis and political science.